Food Security in the Developing World

Food Security in the Developing World

Edited by

Muhammad Khalid Bashir
University of Agriculture
Faisalabad, Pakistan

Steven G.M. Schilizzi
University of Western Australia
Australia

Ghaffar Ali
Shenzhen University
China

WILEY

This edition first published 2024

Registered Office(s)
John Wiley & Sons, Inc., 111 River Street, Hoboken, NJ 07030, USA
John Wiley & Sons Ltd, The Atrium, Southern Gate, Chichester, West Sussex, PO19 8SQ, UK

For details of our global editorial offices, customer services, and more information about Wiley products visit us at www.wiley.com.

Wiley also publishes its books in a variety of electronic formats and by print-on-demand. Some content that appears in standard print versions of this book may not be available in other formats.

Library of Congress Cataloging-in-Publication Data
Names: Bashir, Muhammad Khalid, editor. | Schilizzi, Steven G.M., editor. | Ali, Ghaffar, editor.
Title: Food security in the developing world / edited by Muhammad Khalid Bashir, Steven G.M. Schilizzi, Ghaffar Ali.
Description: Hoboken, NJ : Wiley, 2024. | Includes index.
Identifiers: LCCN 2023040109 (print) | LCCN 2023040110 (ebook) | ISBN 9781119265108 (hardback) | ISBN 9781119265160 (adobe pdf) | ISBN 9781119265177 (epub)
Subjects: LCSH: Food security–Developing countries.
Classification: LCC HD9018.D44 F66564 2024 (print) | LCC HD9018.D44 (ebook) | DDC 338.1/91724–dc23/eng/20240102
LC record available at https://lccn.loc.gov/2023040109
LC ebook record available at https://lccn.loc.gov/2023040110

Cover Design: Wiley
Cover Image: © OlgaKlochanko/Adobe Stock Photos

Set in 9.5/12.5pt STIXTwoText by Straive, Pondicherry, India

M WEP260371 310124

Contents

About the Editors

Dr. Muhammad Khalid Bashir currently holds the position of Associate Professor at the Institute of Agricultural and Resource Economics, University of Agriculture, Faisalabad. His commitment to academia and research extends beyond this role, as he also serves as the co-chair of the Policy, Advocacy, and Outreach thematic division in the Pak-Korea Nutrition Center (PKNC).

Dr. Bashir's academic journey has been marked by significant accomplishments. He completed his PhD at the University of Western Australia, Australia, and his research expertise encompasses critical areas such as food security, poverty, and sustainable rural development. Throughout his career, he has secured numerous research grants from both national and international funding agencies, focusing on improving food availability and addressing food security challenges.

An accomplished scholar, Dr. Bashir has made substantial contributions to the field of agricultural and resource economics. He has authored over 70 research papers that have been published in esteemed national and international journals. His dedication to mentoring the next generation of scholars is evident in his supervision of over 70 students pursuing MS, MPhil, and MSc (Hons) degrees. Notably, three students have successfully completed their PhD studies under his guidance, while five others are currently under his mentorship.

Dr. Bashir's influence extends beyond the borders of Pakistan, as he has co-supervised PhD students from prestigious institutions such as Huazhong Agricultural University in Wuhan, China, and Universiti Putra Malaysia. His exceptional contributions have earned him 18 national and international awards, including recognition for his teaching excellence in 2015, the Best Research Paper Award from the Higher Education Commission of Pakistan in 2016, and the Best Young Researcher Award in 2019. His remarkable dedication to research is further exemplified by the Research

Productivity Award from the Pakistan Council for Science and Technology in 2016, Publons Reviewer Award in 2018, and Outstanding Contribution in Reviewing in the same year. Dr. Bashir's work has been acknowledged with Best Conference Paper awards in 2018 and 2019. Additionally, the University has nominated him for prestigious UNESCO and Alkharzmi awards in 2017 and 2019, respectively, underlining his significant contributions to the field of agricultural and resource economics.

Dr. Steven G.M. Schilizzi is a distinguished academic with a diverse educational background spanning multiple disciplines. Earning his PhD in Economics with a focus on Agriculture, Resources, and Energy in France, Steven is also an alumnus of the Ecole Normale Supérieure (Ulm) in Paris, where he immersed himself in humanities, classics, and languages. Before this, he achieved a Baccalaureate in Math and Physics, a master's degree in Sociology, and a bachelor's in international law.

Early in his career, while affiliated with France's National Institute of Agriculture Sciences in Montpellier (INRA), Steven cultivated rich, on-ground experience across regions such as South-East Asia, central Africa, and Central America. His association with UWA later saw him enhancing his expertise in Vietnam by spearheading various research initiatives and launching collaborative courses between UWA and Vietnamese institutions.

Currently a Professor of Environmental and Agricultural Economics at UWA's School of Agriculture and Environment, Steven's research gravitates toward exploring the confluence of social norms and economic incentives in prompting private landholders to offer environmental services, encompassing areas like biodiversity conservation and pollution control. He is deeply involved in understanding the intricate balance of equity in environmental policies, their trade-offs, and their intersections with other policy areas. Over the years, Steven has shown a penchant for economic lab experiments, intertwining with the realms of behavioral economics.

An accomplished author, Steven has contributed to over 100 scientific publications, penned 20 book chapters, and authored 4 books, one notably being "Natural Resources and Environmental Justice." Steven's mentorship extends to PhD, Masters, and Honors students from varied disciplines, supervising approximately 20 PhD candidates throughout his career. His cross-disciplinary approach in research and teaching is evident, with a key focus on the symbiotic relationship between biological–geographical systems and socio-economic determinants.

Dr. Ghaffar Ali is currently serving as Associate Professor at the College of Management, Shenzhen University, China. His scholarly endeavors have not only garnered him acclaim on a global scale but have also firmly established him as a prominent figure in the field of academia. This distinction is further affirmed by his ranking among the Top 2% of researchers worldwide, as per a recent evaluation conducted by Stanford University in the United States. Moreover, Dr. Ghaffar Ali's unwavering commitment to the advancement of research is underscored by his inclusion among the Top 1% of peer reviewers globally, a distinction conferred by Web of Science through Publons.

In addition to his commendable research accomplishments, Dr. Ali actively engages with the scholarly community, taking on roles as an Editor and serving on the Editorial Boards of four distinguished international journals. His dedication to shaping and enriching academic discourse within his areas of expertise is further manifested by his affiliation with several prominent academic societies. Notable among these affiliations are his memberships in esteemed organizations such as the Institute of Electrical and Electronics Engineers (IEEE), specifically within the IEEE's Energy & Power division, the International Society for Data Science and Statistics (ISDS) in Japan, and the Chinese Economic Society in the United States.

The breadth and depth of Dr. Ali's research interests span a wide spectrum of critical subjects within the domains of economics and energy. He has made substantial contributions to a diverse range of topics, including but not limited to environmental economics, energy economics, food economics, energy policy, renewable energy transition, green building practices, and climate change policy. His work not only contributes significantly to the expansion of knowledge within these domains but also has the potential to shape and influence policy decisions that bear profound implications for environmental sustainability, energy resource management, and economic dynamics. Dr. Ali's multifaceted contributions, coupled with his global recognition in the realms of research and academia, underscore his pivotal role in addressing some of the most pressing challenges of our time. His dedication to advancing knowledge and effecting positive change makes him an indispensable asset to the academic and research communities.

List of Contributors

Namrah Abid
National Institute of Food Science and Technology
University of Agriculture
Faisalabad
Pakistan

Muhammad Abid Bashir
Department of Surgery, Rabigh
Faculty of Medicine
King Abdul Aziz University
Jeddah
Saudi Arabia

Burhan Ahmad
Institute of Business Management Sciences
University of Agriculture
Faisalabad
Pakistan

Ghaffar Ali
College of Management
Shenzhen University
Shenzhen
China

Muhammad Sohail Amjad Makhdum
Department of Economics
Government College University
Faisalabad
Pakistan

Hammad Badar
Institute of Business Management Sciences
University of Agriculture
Faisalabad
Pakistan

Muhammad Hamid Bashir
Department of Entomology
University of Agriculture
Faisalabad
Pakistan

Muhammad Khalid Bashir
Institute of Agricultural and Resource Economics
University of Agriculture
Faisalabad
Pakistan

Muhammad Umar Farrukh
Department of Economics
Government College Women University
Sialkot
Pakistan

Ole Gjølberg
NMBU School of Economics and Business
Norwegian University of Life Sciences (NMBU)
Aas
Norway

Muhammad Hafeez
Institute of Business Management Sciences
University of Agriculture
Faisalabad
Pakistan

Sarfraz Hassan
Institute of Agricultural and Resource Economics
University of Agriculture
Faisalabad
Pakistan

Sabyan Faris Honey
Regional Bioscience Centre
CABI
Rawalpindi
Pakistan

Muhammad Asif Kamran
Institute of Agricultural and Resource Economics, Faculty of Social Sciences
University of Agriculture
Faisalabad
Pakistan

Rakhshanda Kousar
Faculty of Social Sciences
Institute of Agricultural and Resource Economics
University of Agriculture
Faisalabad
Pakistan

Mubashir Mehdi
Department of Agri-Business and Applied Economics
MNS University of Agriculture
Multan
Pakistan

Khalid Mushtaq
Faculty of Social Sciences
Institute of Agricultural and Resource Economics
University of Agriculture
Faisalabad
Pakistan

Tahira Sadaf
Institute of Agricultural and Resource Economics
University of Agriculture
Faisalabad
Pakistan

Kainat Sarwar
National Institute of Food Science and Technology
University of Agriculture
Faisalabad
Pakistan

Steven G.M. Schilizzi
Department of Agricultural and Resource Economics, UWA School of Agriculture and Environment
The University of Western Australia
Crawley, Perth
Australia

Sana Sehar
Institute of Agricultural and Resource Economics
University of Agriculture
Faisalabad
Pakistan

Mian Kamran Sharif
National Institute of Food Science and Technology
University of Agriculture
Faisalabad
Pakistan

Raza Ullah
Institute of Agricultural and Resource Economics, Faculty of Social Sciences
University of Agriculture
Faisalabad
Pakistan

Farhad Zulfiqar
Agribusiness Management
Department of Food
Agriculture and Bioresources
School of Environment
Resources and Development
Asian Institute of Technology (AIT)
Klong Luang
Pathum Thani
Thailand

Preface

Food is a universal need, the very sustenance of life, and yet, access to it remains a privilege for many, which excludes too many fellow humans. The phrase "food security" is not new, but in today's fast-paced, globalized world, it is more relevant than ever. The book, *Food Security in the Developing World*, delves into the complexities of this significant subject, shedding light on the myriad dimensions that are often overlooked in conventional discussions.

The developing world, with its vast landscapes, diverse cultures, and varied agricultural practices, is uniquely positioned in the discourse on food security. While these nations boast an abundance of natural resources and agricultural potential, they often grapple with paradoxes of plenty: hunger amidst abundant harvests, malnutrition in regions of food surplus, and economic disparities that can turn food from a basic right into a luxury.

This book is not just about the scarcity of food. It addresses the nuanced tapestry of socioeconomic policies, historical contexts, climatic challenges, agricultural practices, and even deeply ingrained cultural paradigms that influence food production, distribution, and consumption in the developing world. We will take our readers on a journey across continents, looking closely at real-life narratives, innovative farming practices, policy challenges, and community-led solutions that are making a difference.

Our objective is twofold: First, to provide readers with a comprehensive understanding of the intricate web of factors that influence food security in developing nations. And second, to inspire action and dialogue among policymakers, activists, farmers, and consumers. We believe that a collaborative, informed approach is key to finding sustainable solutions to food insecurity.

As you delve into these pages, we hope you gain insights not just about the challenges but also about the resilience and innovation inherent in the developing world. While the road to global food security is long and fraught with challenges, the stories and analysis in this book aim to illuminate the path and inspire change.

To a future where every individual, irrespective of where they live, has access to nutritious food – that is the vision this book holds close to its heart.

Thank you for joining us on this journey.

Muhammad Khalid Bashir,
Steven G.M. Schilizzi, and Ghaffar Ali

1

Introduction to the Issue: Food Security

Muhammad Khalid Bashir[1], Steven G.M. Schilizzi[2], and Muhammad Umar Farrukh[3]

[1] *Institute of Agricultural and Resource Economics, University of Agriculture, Faisalabad, Pakistan*
[2] *Department of Agricultural and Resource Economics, UWA School of Agriculture and Environment, The University of Western Australia, Crawley, Perth, Australia*
[3] *Department of Economics, Government College Women University, Sialkot, Pakistan*

Even though the world is producing enough food, about 10.4% (800 million) of its population is malnourished (FAO 2022). Developing countries are home to about 98% of the world's undernourished people. These poor people are unable to afford adequate diets, which is the biggest hurdle for achieving the goal of "zero hunger." The situation has been worsened by the COVID-19 pandemic. The dynamics of food security are multifaceted and need special attention. Developing countries are similar in general characteristics (low levels of growth in national income, low levels of per capita growth, highly skewed patterns of income distribution, masses suffering from absolute poverty, ill health and malnutrition, low levels of literacy). It is time to understand the dynamics of "food security" in these countries and what is driving them. A simple but meticulous approach will help in this regard, building on the background and deep insights provided in a more general setting by Acemoglu and Robinson's 2012 book, "Why Nations Fail."

Food security became an issue for academic study in the 1970s (Bashir and Schilizzi 2013). Since then, it has attracted the attention of researchers and policymakers. Initially, food security was defined as the ability of nations to consistently meet their aggregate food needs, especially cereals. Then the influential work of Sen (1981) changed the course of food security focus

Food Security in the Developing World, First Edition. Edited by Muhammad Khalid Bashir, Steven G.M. Schilizzi, and Ghaffar Ali.

from sufficiency (availability) to access. According to him, individuals' ability to access food depends on their entitlements, i.e. access to resources, markets, technology, and social networks. During the 1980s, the focus further narrowed down to the importance of nutritional security. The focus further shifted to market-oriented research in the 1990s. During the 2010s, the challenges facing agricultural productivity under global climate change and limited farmer resources have again modified the focus by linking it to vulnerability analysis.

Three core areas, food availability, accessibility, and utilization, evolved from the well-known and accepted definition of food security, i.e. *food that is always available; that all persons have means of access to it; that it is nutritionally adequate in terms of quantity, quality, and variety; and that it is acceptable within the given culture. Only when all these conditions are in place can a population be considered food secure* (FAO 2004). Understanding these core areas in depth will help us understand the phenomenon of food insecurity and better inform the policy process in developing countries. This book is organized into nine chapters, focusing on the following challenges:

- The second chapter focuses on food availability levels and general policy practices to maintain food stocks in developing countries. Food availability depends not only on the general food supply but also on the policy environment enacted by a country and, in particular, on the incentive structures in place. A favorable environment ensures stability and provides agricultural infrastructure for food supply and for maintaining food stocks. Food availability has therefore been measured in terms of sufficiency of supply and ease of access. This chapter starts with a discussion on building the concept of food security and food availability, followed by the status of food availability in terms of the policy environment across the globe. Sufficiency of supply is discussed in reference to food consumption in terms of kilocalories per day per capita. Thereafter, the relationship between food supply and agricultural infrastructure is explored. A comparison of South Asian, Southeast Asian, and global food stocks follows. The food stocks help in the case of food shortages, lean production seasons, and price volatility. These stocks help maintain a steady food supply and stable prices. A detailed discussion on stock maintenance policies and incentives is presented to identify the rationale and derive policy lessons.
- The third chapter examines the status and economic significance of postharvest losses (PHLs) and discusses the case of fruits and vegetables in Pakistan. PHLs refer to the measurable loss in food quantity and quality from harvest to its final consumption. Losses in food quality refer to the

reduction in both (objective) nutrient contents and (subjective) acceptability by the consumer. These are more common in developed countries (Kader 2002). Losses in quantity refer to the reduction in the amount of product. These are more common in developing countries due to poor sanitation and handling facilities (Kitinoja and Gorny 1999). Therefore, interventions in reducing PHL are necessary to address the key challenge of food security. Postharvest value addition is also considered as an integral part of all strategies aiming to combat food security for a growing population.

- The fourth chapter links climate change to agricultural productivity. Climate change at the global level can affect agricultural systems at the regional and local levels by impacting ecological dynamics, especially when greater and more frequent flood disasters directly affect agricultural production, with immediate consequences on economic and human well-being.
- The fifth chapter presents a bird's-eye view of agricultural revolutions for ensuring food security and sustaining the livelihood of masses, particularly in developing countries. Several agricultural revolutions have happened in human history, including "the Neolithic Revolution," the West-European Technological Revolution, and the North American Mechanization Revolution. However, the most recent and influential agricultural revolution was the Green Revolution of the 1960s. It featured the advent of science and technology, like plant breeding, genetic engineering, and the use of hybrid varieties of grains, fertilizers, pesticides, machines, and improved irrigation systems. In the post-Green Revolution Era, however, nations improved their R&D priorities by incorporating environmental concerns, poverty reduction, and sustainable resource management into farming systems, which may be considered as "a second agricultural revolution."
- The sixth chapter examines the dynamics of income and income redistribution in developing countries. Rising trends of income inequality in the world are cause for much concern, as they can lead to many other economic and social inequalities, each creating yet more challenges. This income gap has in recent decades grown mostly in developed economies; in developing and emerging economies this trend is mixed. The development of any economy depends not only on the total national income but also on the distribution of income. Indeed, the pattern of income distribution over the years has been a major factor in explaining the level of economic growth and development of any economy. The chapter also describes the subject matter with the help of real-world case studies, taken mostly from the Organization for Economic Cooperation and Development (OECD) and from Latin American countries.

- The seventh chapter examines the market availability of food. Food security has recently captured the imagination of public policy planners, thinkers, academia, and industry because of the rapid growth in the world population, which is expected to exceed nine billion in 2050. To meet this challenge, adequate availability of markets for producers and consumers is considered important. Markets facilitate the exchange of goods and services essential for ensuring a country's food security. Other main market functions include price determination, resource allocation, policy implementation, value addition, and employment generation. Markets in developing countries involve different stakeholders, such as producers, consumers, traders, trade supporters, and policy planners, who operate to achieve their own goals. Agricultural producers have the option to directly market their produce, usually through farmers' markets. However, most producers in developing countries opt for nondirect marketing options and sell their produce to traders, exporters, processors, or retailers, who then pass on this produce to consumers. Consumers buy their food requirements either from traditional or modern retailers. The performance of markets in developing countries cannot be termed optimal, mainly because of inequitable returns, the predominance of intermediaries, transportation and storage problems, credit constraints, lack of modern marketing and technical knowledge, defective measurement equipment, and lack of organized markets. To overcome the challenges markets face in developing countries, needed policy measures may include promotion of an entrepreneurial mindset, market governance, institutional innovations, diversification, standards, and certification, and greater use of information and communication technology (ICT).
- The eighth chapter covers the developing world's food (in)security challenges using a multifaceted approach. Prices send important signals to producers and consumers. Based on these signals, they make decisions. Price volatility poses difficulties in the efficient allocation of resources both for producers and consumers but may also do so for various stakeholders involved in the supply chain. Consequently, policymakers aim to formulate policies to better control price volatility. Repeated episodes of volatility may lead to food insecurity because of decisions responsible for the inefficient allocation of resources. As a basis for developing good policies, it is important to measure and identify the causes of price volatility and to devise policies for controlling it. Different methods are formulated to estimate conditional and unconditional volatilities. Standard deviations and coefficients of variation are generally used to measure unconditional volatilities, while univariate and multivariate GARCH models are widely

used to model conditional volatility. Price volatility, particularly of food commodities, has implications for all the supply chain actors, market integration and food security and attracts the attention of policymakers. A case study of price volatility in Pakistan's rice markets is an effort to address the aforementioned features and is also presented here.

- The ninth chapter links the issue of food safety to sanitation. Safe food is a basic human right, which means that food should be free of any hazards and contaminants that pose a risk to human health. With advances in technology and globalization, food safety systems have come into existence, but at the same time new challenges for food safety have arisen, like climate change, new adulterants, use of additives, etc. Food safety requires effort at every step, from farm to plate. Sanitation and hygiene are one of the most important factors that greatly impact food safety. Food safety without sanitation is not possible. Hence, the focus should be on identifying the conditions under which the principles of sanitation and hygiene can be followed by everyone related to the food supply chain, either as producer, handler, or consumer. International organizations, governments, and stakeholders in the food industry should join hands to promote research, laws, awareness, and knowledge regarding food safety and sanitation so that the global burden of foodborne diseases around the world can be decreased. To sum up, the remaining chapters investigate the following key drivers of food (in)security: food availability; post-harvest losses; the impacts of climate change on agricultural productivity; the history of how food security was addressed over time; and the ways in which income and wealth distribution, market governance, price volatility and sanitation and hygiene all affect food security.

References

Acemoglu, D. and Robinson, J.A. (2012). *Why Nations Fail. The Origins of Power, Prosperity and Poverty*. New York: Crown Publishers.

Bashir, M.K. and Schilizzi, S. (2013). Determinants of rural household food security: a comparative analysis of African and Asian studies. *Journal of the Science of Food and Agriculture* 93: 1251–1258.

FAO (2004). *FAO Statistics*. Rome: Food and Agriculture Organization.

FAO (2022). *The State of Food Security and Nutrition in the World: Transforming Food Systems for Food Security, Improved Nutrition, and Affordable Healthy Diets For All*. Rome: Food and Agriculture Organization.

Kader, A.A. (2002). *Postharvest Technology of Horticultural Crops*, 3e. Oakland: University of California, Agriculture andNaturalResources. Publ3311.

Kitinoja, L. and Gorny, J.R. (1999). *Postharvest Technology for Small-scale Produce Marketers: Economic Opportunities, Quality and Food Safety*. University of California Postharvest Horticulture. Series No. 21.

Sen, A. (1981). *Poverty and Famines: An Essay on Entitlement and Deprivation*. Oxford: Clarendon Press.

2

Food Availability: Stock Maintenance Policies

Farhad Zulfiqar[1], *Raza Ullah*[2], *and Muhammad Asif Kamran*[2]

[1] *Agribusiness Management, Department of Food, Agriculture and Bioresources, School of Environment, Resources and Development, Asian Institute of Technology (AIT), Klong Luang, Pathum Thani, Thailand*
[2] *Institute of Agricultural and Resource Economics, Faculty of Social Sciences, University of Agriculture, Faisalabad, Pakistan*

This chapter focuses on food availability levels and general policy practices to maintain food stocks in developing countries. The food availability depends not only on the general food supply but also on the policy environment enacted by a country. A favorable environment ensures stability and provides agricultural infrastructure for food supply and maintaining food stocks. Therefore, food availability was measured in terms of sufficiency of supply and ease of access. This chapter starts with a discussion on building the concept for food security and food availability followed by the status of food availability in terms of the policy environment across the globe. Then, sufficiency of supply is discussed concerning food consumption in terms of kilo calorie per day per capita. Thereafter, the relationship between food supply and agricultural infrastructure is explored. A comparison of South Asian, Southeast Asian and global food stocks follows. The food stocks help in the case of food shortages, lean production seasons, and price volatility. Thus, these stocks help in maintaining a steady food supply and price stabilization. A detailed discussion on stock maintenance policies is presented to identify the rationale and policy lessons.

2.1 Food Security and Food Availability

Food security is a dynamic concept and has seen various dimensions added to its definition (FAO 2006). The most widely agreed definition from the World Food Summit 1996 stresses the achievement of dimensions, including food

Food Security in the Developing World, First Edition. Edited by Muhammad Khalid Bashir, Steven G.M. Schilizzi, and Ghaffar Ali.

availability, access, utilization, and stability (Bashir and Schilizzi 2015; FAO 1996, 2006, 2008). The underachievement of food security may lead to any of the two types of food insecurity: chronic food insecurity or transitory food insecurity (FAO 2008). However, there is another possibility of people being food-insecure—it is probable that people face seasonal food insecurity due to cycles of production (FAO 2008) or seasonal demand changes (Hussain et al. 2014). Agricultural production is extremely dependent on the external environment, and staple grains' production is highly seasonal. Therefore, the food availability across seasons requires food stocks. Thus, stock maintenance policies form an important basis of overall food security policies of a country.

2.2 Food Availability

Food availability is an intrinsic component of food security, which also includes various other dimensions, including food affordability, food quality, and food safety (EIU 2016). The food availability needs to be considered as a dynamic concept because of changing food demand and food production. Demand changes due to improved income level and increasing population necessitate supply enhancements (Zulfiqar and Hussain 2014), which can be used for stocks. Food stocks are necessary to act as a buffer in the face of various risks to food availability. These include climate change-related risks and changes in government regulations that risk stability in food availability and food prices (EIU 2016).

However, the availability of food hinges not only on the food supply but also on the ease of access to food. Therefore, food availability should be seen as a comprehensive package of food supply, access, distribution, and structural traits of a country's capacity (EIU 2016). The structural traits include, but not limited to, good governance and policies for sufficient and affordable food, including fair distribution. These policies must also include stock maintenance policies because of the importance of stocks for supply and price stabilization. The expansion of food availability definition by the Food and Agriculture Organization (FAO), which defines food availability in terms of sufficiency of good quality food through domestic or imported food (FAO 2006), into food supply as well as food access provides a clearer overview of food stocks needed and the appropriate policies for it.

The food availability in this chapter is discussed in length in terms of food supply and the overall policy environment ensuring food access. An index created by the Economist Intelligence Unit captures the same dimensions as required for the explanation of food availability envisioned here. This food availability index measures the status of food availability using various diverse indicators. These eight indicators included "sufficiency of supply, public expenditure on agricultural research and development (R&D),

agricultural infrastructure, volatility of agricultural production, political stability risk, corruption, urban absorption capacity, and food loss" (EIU 2016). Therefore, this indicator was used to elaborate the status of food availability and its numerous dimensions.

2.2.1 Food Availability Status

The food availability status, measured in terms of the aforementioned eight indicators, showed the appropriateness of the policy environment. The countries were ranked on a scale of 0, for worst, and 100, for best policy environment for supply and access. This means a higher score entails a better environment.

It can be deduced from Figure 2.1 that most countries in North America and Europe had the best policy environment to ensure food availability. However, there were some countries in Australia and Asia too that showed good environment, including Australia, Japan, Malaysia, and Singapore. There were some similarities in all of these countries showing best scores which showed that an appropriate environment (referring to favorable policy environment) is necessary for food availability for ensuring food security. These countries have advanced agriculture, fewer structural problems, and higher agricultural hard and soft infrastructure (physical infrastructure, and research and innovation).

A majority of countries with low scores are in Africa. There are several reasons for this tendency. The biggest problem in these countries is the physical agricultural infrastructure (EIU 2016). The environment essential for ensuring sufficient food availability is not available in these countries. This low infrastructure is also linked to low investments, which is a result of capacity constraints of these countries. These are mostly low-income countries. The countries having moderate environment scoring between 48% and 57% include major population centers and need major developments if we are to increase global food availability. These countries include India, Russia, Pakistan, and Indonesia.

Thus, it is appropriate to conclude from food availability variations across countries (Figure 2.1) that the existence of proper policy environment has an important link with food availability across continents.

2.2.2 Demand and Supply—Case Studies from Asia and Sub-Saharan Africa

The overall food availability should be seen as having individual demand and supply components. The analysis of food availability from this perspective provides important insights into domestic policies and practices. A country with high domestic demand for a food product but less supply is

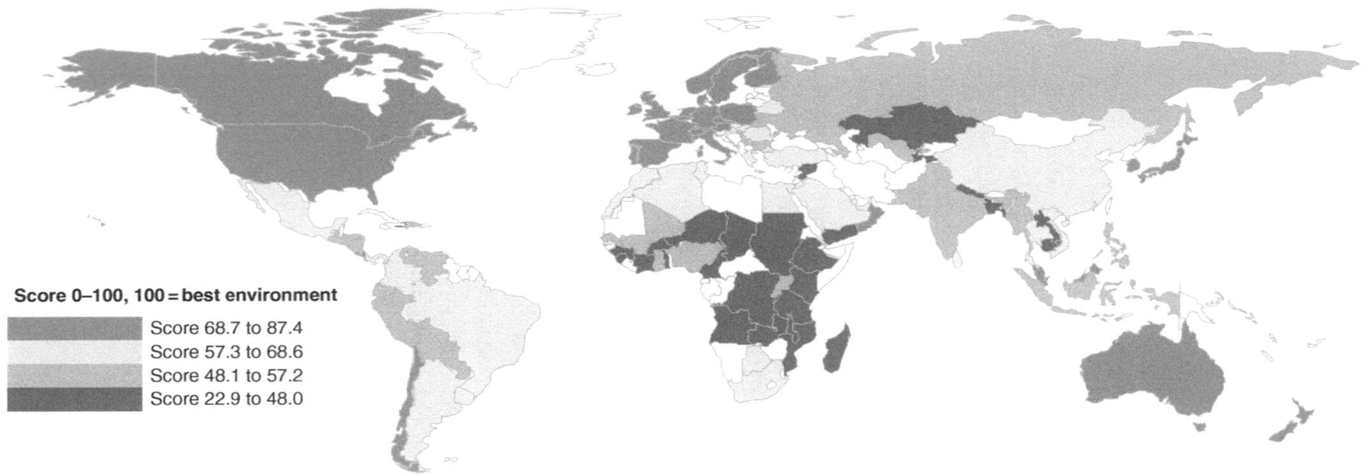

Figure 2.1 Food availability. *Source:* Adapted from the FAO and WFP (EIU 2016).

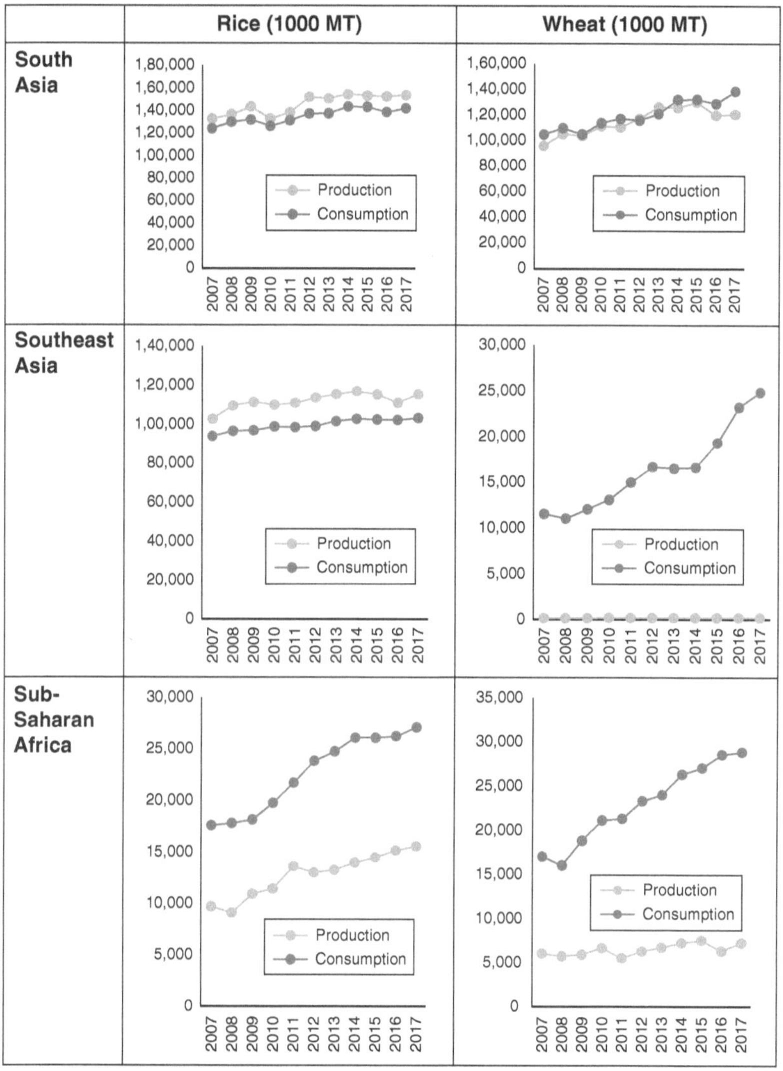

Figure 2.2 Domestic production and consumption comparisons of Asia and sub-Saharan Africa. *Source:* Data from USDA (2017).

expected to enable policies conducive for increased domestic production or liberal import regime. An appropriate policy environment is, therefore, essential for ensuring sufficient food availability in a country.

Generally, South Asia and Southeast Asia were self-sufficient in rice availability during 2007–2017 (Figure 2.2). The domestic production is more than covered for domestic consumption. However, sub-Saharan Africa faced a huge supply gap which further increased during the considered time

(Figure 2.2). An appropriate policy environment will ensure that surplus rice in South Asia and Southeast Asia is exported to earn foreign exchange and demand gap in sub-Saharan Africa will be met through imports or increased domestic production. However, low scores on the policy environment (Figure 2.1) showcase an impending hurdle to achieve this goal. This provides a clear framework for bridging the demand gap in sub-Saharan Africa through improvement in the policy environment.

In the case of wheat, only South Asia has been able to satisfy demand through domestic production during 2007–2017 (Figure 2.2). Wheat demand has seen an exponential growth in Southeast Asia and sub-Saharan Africa in recent years (Figure 2.2). However, the domestic supply has either remained flat (Southeast Asia) or increased very slowly (sub-Saharan Africa). This calls for appropriate policy environment in South Asia to remain wheat self-sufficient and in other two regions to provide an enabling environment for imports or domestic production promotion. The trade policies in the latter regions are very relevant in ensuring food availability.

2.2.3 Sufficiency of Supply

The sufficiency of supply (EIU 2016) was expressed here as an indicator of food availability in terms of average calorie intake as kcal/capita/day, including food aid. It is an indicator of net food consumption arising from domestic production and imports while subtracting the food exports. The countries were ranked from 1 to 113 based on average calorie intake with 1 referring to highest sufficiency and 113 as lowest (Table 2.1).

Table 2.1 Country rankings for sufficiency of supply (113 countries).

Top 10 scores			Bottom 10 scores		
1	Belgium	100.0	104	Bangladesh	26.7
2	Austria	99.7	105	Madagascar	26.2
3	Turkey	96.1	106	Cambodia	25.4
4	United States	94.7	107	Malawi	22.7
5	Israel	94.0	108	Zambia	22.0
6	Ireland	93.0	109	Mozambique	20.9
=7	Germany	91.2	110	Congo (Dem. Rep.)	17.9
=7	Italy	91.2	111	Ethiopia	15.7
9	France	90.7	112	Chad	14.9
10	Switzerland	89.4	113	Burundi	0.0

A composite indicator that measures the availability of food through the food supply in kcal/capita/day and levels of food aid.
Source: Adapted from EIU calculation based on data from the FAO and WFP (EIU 2016).

The developed countries, exclusively high-income countries, ranked highest in supply sufficiency indicator, while the worst-performing countries are either from Asia Pacific, South America, or Africa. The comparison of Table 2.1 and Figure 2.1 highlights important policy directions. The countries having good policy environment (Figure 2.1) also ranked high in terms of sufficiency of supply. While the majority of African countries with unfavorable policy environment also ranked lowest in sufficiency of supply. This confirms the notion that the overall environment affects the food sufficiency across regions.

The food supply sufficiency requires a good environment whereby public and private entities can work together to ensure food access and storage across seasons. The public investment in agricultural research and development also drives supply levels as it provides much-needed technology for public and private use. Thus, it is necessary to have sufficient political will to ensure sufficiency of supply.

2.2.4 Sufficiency of Supply and Agricultural Infrastructure

The agricultural infrastructure needed for the year-round supply of food includes storage facilities along with land and sea transportation services. The prioritization of physical infrastructure development, such as agricultural storage and transportation, enhances a country's capacity to safeguard sufficiency of supply for increasing demand due to growing population (EIU 2016). The development of this physical infrastructure should be carried out through public and private partnerships to sustain the investment and infrastructure capacity for burgeoning future needs.

The emphasis on physical agricultural infrastructure cannot be overstated as there has been a global direct correlation between the availability of physical infrastructure and sufficiency of supply (Figure 2.3).

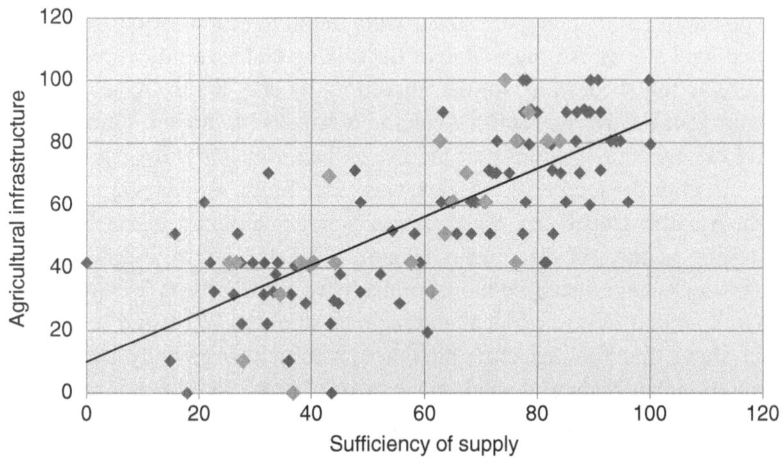

Figure 2.3 Sufficiency of supply vs agricultural infrastructure. *Source:* Adapted from EIU (2016).

Note: The agricultural infrastructure includes all physical infrastructure, including storage, road, and port infrastructure for agriculture. The highlighted countries are from Asia and Pacific.

Figure 2.3 shows that a low agricultural infrastructure availability is correlated with low sufficiency of food supply. The countries in Asia and Pacific are scattered all along the graph, but the direct relation has been true. Also, almost all the countries from Africa were clattered in the lower left corner of Figure 2.3. The scarcity of physical agricultural infrastructure in Africa has led to low sufficiency of supply in the region.

There are two dimensions of the relation between agricultural infrastructure and food supply. First, the low capacity of physical infrastructure results in chronic shortages in the lean production period. Second, low infrastructure leads to low incentives for surplus food production because of limited marketing opportunities leading toward focus on subsistence agriculture and inadequate marketable surplus.

2.2.5 Food Stocks

The food stocks in South Asia, Southeast Asia, and the World are compared here to present a picture of global interconnectedness, despite the varying national interests of countries (Figure 2.4a and b). The world stocks in food grains (wheat plus rice) have been steadily rising since the food crisis of 2007–2008. The stocks suddenly rose across the world and more so in South and Southeast Asia in response to this crisis to protect domestic consumers and to ensure food availability. This was done more extensively in South Asia due to protectionist policies by India and China, resulting in reasonably better managing of the food crisis (Gouel 2013). However, the opportunity cost of these protectionist policies is high (Dorosh 2009).

The South Asian food grain stocks were found to be loosely connected to global food stocks. An increasing or decreasing global trend in stocks was generally followed in a similar direction by the South Asian region (Figure 2.4a). However, South Asia and Southeast Asia were found to be more closely connected in their food stock behavior. The response of both regions, after the food crisis of 2008, was exceedingly similar (Figure 2.4b). Although the global and South Asian stock maintenance can be seen diverging in the last three years (Figure 2.4a), the South and Southeast Asian stocks have displayed more connectivity in these years (Figure 2.4b).

Thus, similar trends in global and regional stocks in South and Southeast Asia show that the stock maintenance is somehow globally connected through global trade and stock mechanism. Therefore, the stock maintenance policies of major trading players affect the global prices and stability.

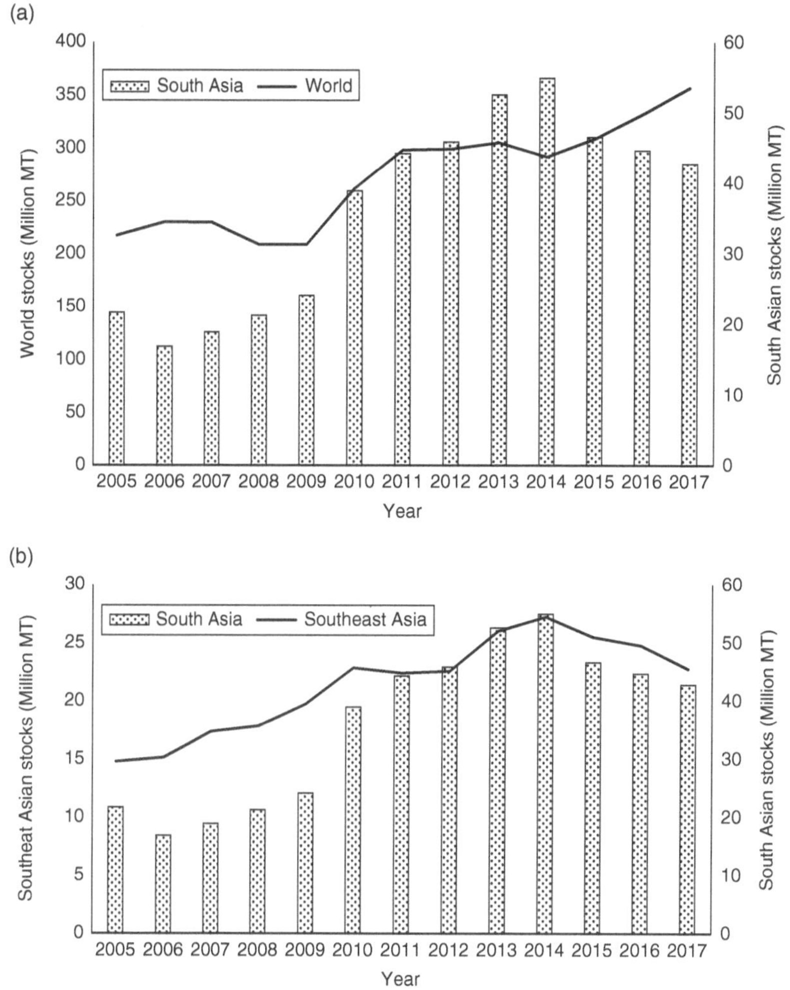

Figure 2.4 (a) Food grain (wheat plus rice) year beginning stocks in South Asia and World. (b) Food grain (wheat plus rice) year beginning stocks in South and Southeast Asia. *Source:* Data from USDA (2017).

2.3 Stock Maintenance Policies

The food availability and subsequent sufficiency of supply across seasons at reasonable prices are ensured in most of the countries through food stocks. The developing countries often resort to more stringent stock maintenance policies to protect consumers and producers from international price volatility and production instability. This is because these countries can ill-afford to pay for the additional costs of importing high-priced food imports in case of shortages. These stock maintenance policies have differing effect on domestic price structure and international prices. For example, the food security policy choices of Pakistan and India during the 2007–2008 food crisis had major spillover effects on neighboring Afghanistan and Bangladesh (Dorosh 2009). The maintenance of food stocks ensures the sufficiency of food supply at affordable prices for domestic consumers in case of production shortfalls, both domestic and international. However, in case of production shortfalls, it may negatively affect international prices if the country is a major exporter of the particular commodity because of export restrictions to maintain food stocks. Also, it may put the domestic producers at a disadvantage as they will not be able to export their commodity at higher international prices. Thus, stock maintenance policies have multiple facets that need to be considered from a policy perspective. The stock maintenance policies of a few major developing countries are discussed in the following section to draw meaningful conclusions about the nature and success of these policies.

2.3.1 Indian Case Study

India is a major player in the international grain market due to its enormous production and consumption levels, and changes in its food demand can significantly affect world prices (del Ninno et al. 2007; Dorosh 2009). The dual objectives of price stabilization and domestic food availability for consumers are met through Food Corporation of India (FCI). The FCI has been successful in achieving these objectives by maintaining price stability for domestic farmers and consumers alike and reducing volatility in markets (McCreary 2012). However, these policies in India may have had negative influence on international markets as India is a major producer and consumer market of grains. The protectionist policies may have contributed to greater international price volatility and higher prices in other countries (Deuss 2014; Dorosh 2009).

The food grain stocks, the staples include wheat and rice, maintained by FCI varied in response to domestic requirements and global changes in supply and prices (Figure 2.5). A study conducted in 2012 concluded that the stock maintenance policies by FCI during 2006–2012 have been successful in price stabilization and incentivizing producers simultaneously (McCreary 2012). This was made possible by stock maintenance policies adopted by India during the international price hike in 2007–2008, whereby the Indian government restricted the exports of wheat and rice (Dorosh 2009; Gouel 2013). This phenomenon can be observed clearly in Figure 2.5 where

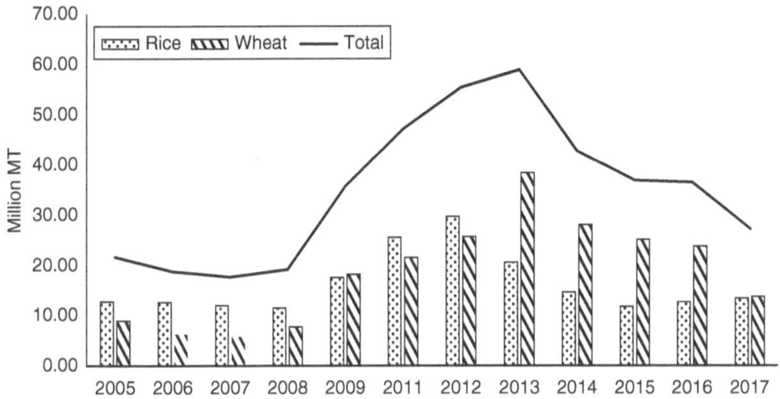

Figure 2.5 Food grain stocks in India. *Source:* Adapted from FCI (2017).

the food stocks started accumulating rapidly due to the export ban as the Indian government prioritized domestic price stability and food availability. It benefited the local populace as the domestic price increase during the food crisis of 2007–2008 was a paltry 7.9% as compared to global price hike of 160% (World-Bank 2010).

Although the stock maintenance policies of India succeeded in stabilizing domestic prices and sufficiency of food supply, they have come with significant local and global costs (Dorosh 2009). The domestic fiscal costs of stock maintenance have been very high and may be better utilized elsewhere, and the global costs were also substantial as the export ban contributed to increase in prices in trading partner countries (Deuss 2014; Gouel 2013). Also, the Indian stock maintenance policies have been muddled since successfully averting the 2007–2008 food crisis—the stocks were not reduced in response to easing of global food crisis resulting in high opportunity cost of public resources (Dorosh 2009).

Thus, it can be concluded from the earlier discussion that the stock maintenance policies have been partially successful in ensuring sufficiency of supply and price stabilization, but they have come at a high cost. The continued accumulation of stocks (Figure 2.5) has persisted despite opportunities for foreign exchange earnings to present themselves without adversely affecting the food security objective of FCI (Gouel 2013). Therefore, the stock maintenance policies of India have left much to be desired after the food crisis of 2007–2008 was averted. Also, a disproportionate increase in food stocks through public procurement has the potential to impede food security due to less market supply of grains (Deuss 2014).

2.3.2 Bangladesh Case Study

It is vital to note that the grain stock maintenance policies of India have an overbearing effect on the prices and policies in Bangladesh because of their high dependence on grain imports from India. However, the stock policies in Bangladesh have moved away from government control toward private sector since the 1990s (Dorosh 2009).

The dynamic nature of food stocks in Bangladesh (Figure 2.6) compared to India (Figure 2.5) can be easily traced to the nature of government policies. The food stocks in Bangladesh have been much more dynamic compared to other Asian countries due to the major involvement of the private sector. The private sector's dominance over food stock imports and maintenance provides Bangladesh more flexibility in managing grain trade. However, it can also lead to domestic price instability in the event of global price volatility. A high Bangladeshi dependence on India for food grain imports has sometimes resulted in higher domestic prices—an Indian rice export ban in 2007–2008 during the global food crisis resulted in a price hike in Bangladesh (Dorosh 2009). Therefore, it can be determined from the earlier discussion that Bangladesh pursues private sector-led stock maintenance policies and the cost of these policies to public exchequer is minimal.

2.3.3 Pakistan Case Study

Pakistan is largely self-sufficient in food grains (Bashir et al. 2012; Hussain and Routray 2012) and maintains a high level of state control over grain prices. This is to make sure that the food prices are within the

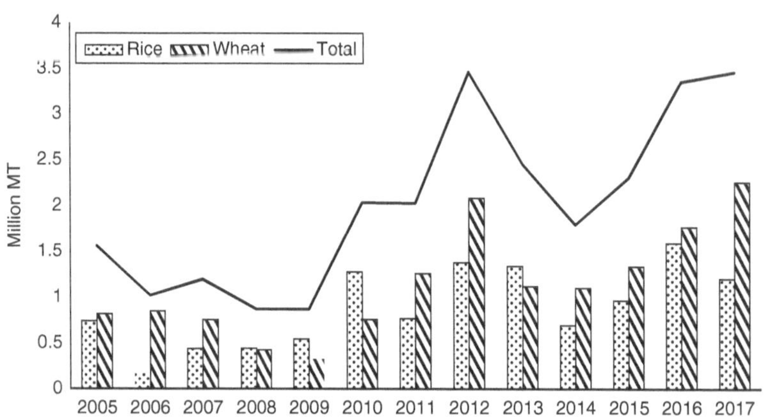

Figure 2.6 Food grain stocks in Bangladesh. *Source:* Data from USDA (2017).

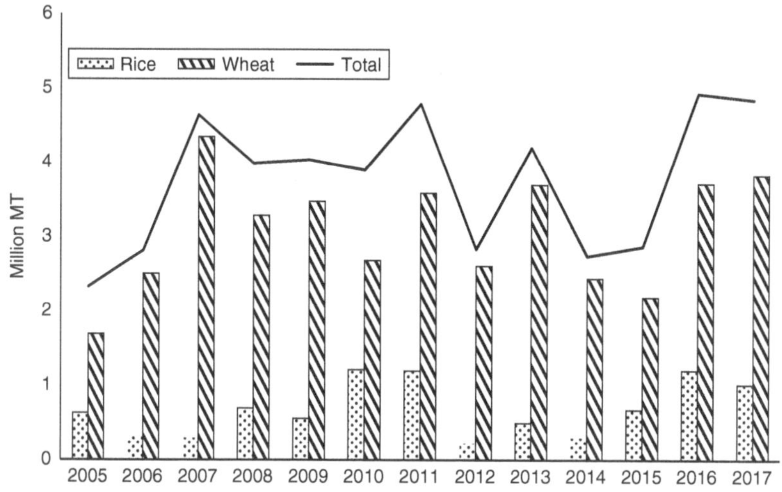

Figure 2.7 Food grain stocks in Pakistan. *Source:* Data from USDA (2017).

reach of consumers and food availability can be ensured at reasonably affordable prices. This is particularly true for wheat—which is major staple food in Pakistan and constitutes majority of food stocks (Figure 2.7).

The grain stocks in Pakistan have varied between 2 and 5 million MT during 2005–2017 (Figure 2.7). The variability in stocks has been due to variations in production and imports to maintain domestic prices. As opposed to Bangladesh, Pakistan's food stocks are managed by public enterprises (Dorosh 2009). Government operates at the market level by purchasing grains at the support prices and distributing them to flour mills at subsidized rates to ensure low consumer prices. However, research has shown that private sector operation is more efficient in ensuring supply at the time of production shortfalls and in maintaining price stability (Dorosh and Salam 2008). Also, stock maintenance by the state carries high fiscal costs (Dorosh 2009), and private sector involvement can helpcontrol these costs (Dorosh and Salam 2008).

Pakistan has often banned grain exports, especially wheat, in the years of production shortfalls to maintain domestic food prices. This has resulted in low domestic prices but caused high prices in neighboring Afghanistan because of its high dependence on imports from Pakistan (Dorosh 2009). The wheat prices in Pakistan and Afghanistan are highly correlated and move together in the long run (Chabot and Dorosh 2007). Thus, it is important to note that, as in the case of Indian policy influence on Bangladesh and Pakistan's on Afghanistan, food policies carry significant domestic fiscal costs along with international price stability ramifications.

2.3.4 Chinese Case Study

The Chinese case study is an interesting example as China considers food stocks information as state secret. Therefore, an estimated amount of stocks has been used here for wheat and rice from the data gathered from USDA (2017). The government provides price support to farmers in order to meet their stocking requirements (Carter et al. 2012). These support prices are steadily raised (Deuss 2014) to provide incentives for more production, resulting in a gradual increase in food grain stocks (Figure 2.8).

The wheat and rice stocks have been steadily increasing in China, more so after the food price crisis of 2007–2008 (Figure 2.8). China, like India and Pakistan, was better able to withstand the global food crisis of 2007–2008 due to interventions in trade and food stocks (Gouel 2013). Perhaps, this success is a result of the combined effect of trade and stock maintenance policies (Deuss 2014). However, this continuous food stock buildup has immense opportunity costs as the Chinese government has accumulated food stocks disregarding any global price fluctuations. Therefore, as pointed out in the earlier discussion throughout this chapter, the policy environment derives the benefits and costs of stock maintenance as well as its success or failure.

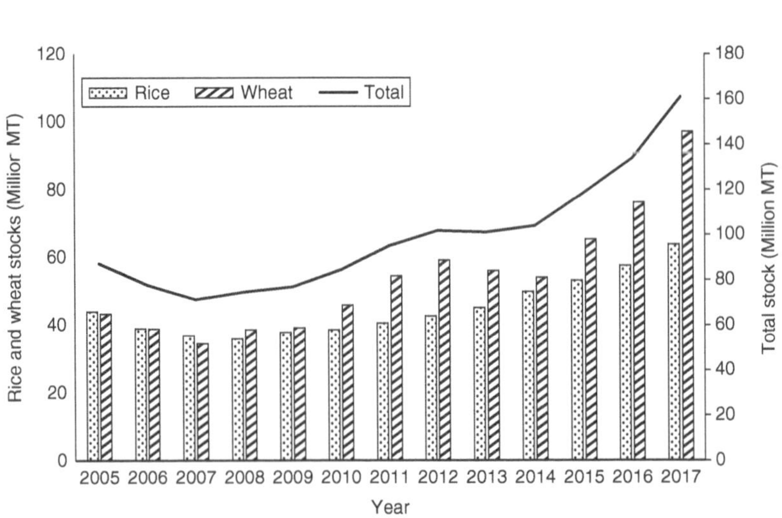

Figure 2.8 Food grain stocks in China (estimated). *Source:* Data from USDA (2017).

2.4 Conclusions

The food availability is one of the critical components of food security. The availability of food depends heavily on the sufficient supply of food and appropriate policies enabling people to have an easy access to it. Food stocks are an important part of any policy to maintain a steady supply of food across seasons and to keep the food prices in check. The findings of this research highlighted that the enabling environment is directly linked to sufficiency of food supply. This environment was well directed in majority of North American and European countries. It needs major overhaul in most of the African countries and some restructuring in major population centers of Asia such as India, Russia, Pakistan, and Indonesia. The food supply sufficiency index, measured in kcal/day/capita, confirmed the link between food supply and policy environment. The African countries ranking low in proper environment were also the ones at the bottom of the list in supply sufficiency among 113 listed countries globally. This is because the food supply sufficiency has a direct correlation with agricultural infrastructure, including transportation and food stocks (Figure 2.3).

Countries have resorted to food stocks since ages, and this is no different today. The food stocks in South and Southeast Asian regions have been rising similar to global trend (Figure 2.4a and b). The stock maintenance policies vary across countries, but with comparable goals of supply sufficiency and price stabilization for consumers and producers alike. The case studies of India, Pakistan, and China present examples of how successful stock maintenance policies can shield countries from global food shortages and price hikes. In India, the objectives of price stabilization and domestic food availability for consumers met through FCI have been successful in maintaining price stability for domestic farmers and consumers alike and reducing volatility in markets. Further, Bangladesh pursues private sector-led food stock maintenance policies, and the cost of these policies to public exchequer is minimal. While, the Pakistan government operates at market level by purchasing grains at the support prices and distributing it to flour mills at subsidized rates to ensure low consumer prices that is relatively less efficient as compared to private sector in ensuring supply at the time of production shortfalls and in maintaining price stability. Like India and Pakistan, China is better able to withstand the global food crisis as a result of the combined effect of trade and stock maintenance policies.

On the contrary, the case studies show that Indian domestic fiscal costs of stock maintenance have very high opportunity cost of public resources and the grain stock maintenance policies have an adverse effect on the prices and policies in Bangladesh because of high dependence on grain imports from India. On the other hand, Pakistan often bans grain exports, especially wheat, in the years of production shortfalls to maintain domestic food prices, which results in low domestic prices but causes high prices in neighboring Afghanistan because of its high dependence on imports from Pakistan. In case of stock policies, Bangladesh has shifted from government control to private sector which leads to domestic price instability in the event of global price volatility; however, Pakistan stock maintenance is controlled by the state which carries high fiscal costs. As in China, continuous food stocks buildup has immense opportunity costs because the Chinese government has accumulated food stocks disregarding any global price fluctuations.

It must be pointed out that these policies carry considerable costs as there are limited public resources. Therefore, a careful balance needs to be found by the policymakers between supply sufficiency and efficiency of public resource use. Nevertheless, these stock maintenance policies work in tandem with trade and other policies. Thus, consistency in policies across the spectrum is needed for desirable results of stock maintenance.

References

Bashir, M.K. and Schilizzi, S. (2015). Food security policy assessment in the Punjab, Pakistan: effectiveness, distortions and their perceptions. *Food Security* 7: 1071–1089. https://doi.org/10.1007/s12571-015-0489-y.

Bashir, M.K., Schilizzi, S., Pandit, R. (2012). Food security and its determinants at the cross roads in Punjab Pakistan, Working Paper 1206, School of Agricultural and Resource Economics, University of Western Australia, Crawley, Australia. 10.22004/ag.econ.126033.

Carter, C.A., Zhong, F., and Zhu, J. (2012). Advances in Chinese agriculture and its global implications. *Applied Economic Perspectives and Policy* 34: 1–36. https://doi.org/10.1093/aepp/ppr047.

Chabot, P. and Dorosh, P.A. (2007). Wheat markets, food aid and food security in Afghanistan. *Food Policy* 32: 334–353. https://doi.org/10.1016/j.foodpol.2006.07.002.

Deuss, A. (2014). Review of stockholding policies. In *OECD Global Forum on Agriculture: Issues in Agricultural Trade Policy, Organization for Economic Cooperation and Development, Agro-food Trade and Markets Division, Trade and Agriculture Directorate*. Paris: OECD.

Dorosh, P.A. (2009). Price stabilization, international trade and national cereal stocks: world price shocks and policy response in South Asia. *Food Security* 1: 137–149. https://doi.org/10.1007/s12571-009-0013-3.

Dorosh, P. and Salam, A. (2008). Wheat markets and price stabilisation in Pakistan: an analysis of policy options. *The Pakistan Development Review* 47: 71–87.

EIU (2016). *Global Food Security Index 2016: An Annual Measure of the State of Global Food Security*. EIU.

FAO (1996). Rome declaration on world food security and world food summit plan of action. *World Food Summit,* Rome (13–17 November 1996).

FAO (2006). *FAO Policy Brief: Food Security* (No. Issue 2). FAO.

FAO (2008). *Introduction to the Basic Concepts of Food Security*. Rome: Food Security Information for Action.

FCI (2017). *Foodgrain Stocks*. India: Food Corporation http://fci.gov.in/ (accessed 07 march 2018).

Gouel, C. (2013). Food price volatility and domestic stabilization policies in developing countries. In: *The Economics of Food Price Volatility* (ed. J.-P. Chavas, D. Hummels, and B. Wright). Chicago: University of Chicago Press.

Hussain, A. and Routray, J.K. (2012). Status and factors of food security in Pakistan. *International Journal of Development. Issues* 11: 164–185. https://doi.org/10.1108/14468951211241146.

Hussain, A., Zulfiqar, F., and Saboor, A. (2014). Changing food patterns across the seasons in rural Pakistan: analysis of food variety, dietary diversity and calorie intake. *Ecology of Food and Nutrition* 53: https://doi.org/10.1080/03670244.2013.792076.

McCreary, I. (2012). *Food Reserves in India*. Paper. Winnipeg, MB: Canadian Foodgrains Bank.

del Ninno, C., Dorosh, P.A., and Subbarao, K. (2007). Food aid, domestic policy and food security: contrasting experiences from South Asia and sub-Saharan Africa. *Food Policy* 32: 413–435. https://doi.org/10.1016/j.foodpol.2006.11.007.

USDA (2017). *Production, Supply and Distribution Online*. Foreign Agricultural Service Off. USDA Estim http://www.fas.usda.gov/psdonline/ (accessed 07 March 2018).

World-Bank (2010). *Food Price Increases in South Asia: National Responses and Regional Dimensions*. Washington, DC.

Zulfiqar, F. and Hussain, A. (2014). Forecasting wheat production gaps to assess the state of future food security in Pakistan. *Journal of Food and Nutritional Disorders* 3: https://doi.org/10.4172/2324-9323.1000146.

3

Postharvest Losses as a Factor of Food Insecurity: The Case of Fruit and Vegetables in Pakistan

Muhammad Hamid Bashir[1] *and Sabyan Faris Honey*[2]

[1] *Department of Entomology, University of Agriculture, Faisalabad, Pakistan*
[2] *Regional Bioscience Centre, CABI, Rawalpindi, Pakistan*

3.1 Introduction

Availability of quality and quantity of food is considered as one of the main focuses for human beings apart from having an appropriate shelter for living, better education, and proper health care. In all discussions related to human life, and especially to poverty alleviation, food security has always been a central focus of discussion. Food security has driven scientists and policymakers all over the world to think about ways to meet the dietary requirements of the growing population. Food security is the situation where all people, all the time, have economic, social, and physical access to sufficient, safe, and healthy food to meet their daily dietary needs and food preferences to have an active and healthy life (FAO 2002).

During the 1970s, availability of food was the sole component for addressing food security that was directly related to food production systems of the country, for linking the level of awareness and technological adaptations among producers and processors engaged in food supply chains. Later, in the 1980s, the concepts of access, utilization, and stability were elaborated for food security for the coming eras. Furthermore, the green revolution provided some relief for addressing food shortages in the world. But it became clear in a very short time that transportation and access to available food are the most determining factors for achieving food security for the people. Thus, policies were shifted by taking on board the

Food Security in the Developing World, First Edition. Edited by Muhammad Khalid Bashir, Steven G.M. Schilizzi, and Ghaffar Ali.

concept of access that refers to the ability of people to obtain food physically and economically.

In recent years, rapid growth in population and increased individual income growth have exacerbated the importance of food security in the world. This is especially true of Asia, which covers most of the developing countries and is the most populated region, occupying half of the world's total population. Based on populations' increasing trend, it is expected that world population will reach 9.3 billion by 2050 (UN 2010), which would clearly exacerbate the issue of global food security in coming decades. According to the Food and Agriculture Organization, there would be an increase of 0.29% in yearly food consumption per person through 2030 (FAO 2006). Furthermore, economic development would also tend to shift the pattern of food consumption from caloric daily intake to quality foods, including processed food, meat, fruit, and vegetables to meet their dietary and social requirements (Table 3.1). This imposes key challenges on developing economies for meeting the people's evolving dietary needs, because it requires a complex global food trade system that has a direct influence on food prices, given consumer demand. This is the reason why, despite rapid economic growth and high living standards in recent years, most of the Asian continent is considered as a hotspot for food insecurity and unequal distribution of food. Malnutrition is also an important challenge for Asia, where 81% of undernourished children are located (World Bank 2009). In Asia, India is one of the most undernourished countries in the world, where 44% of children below the age of 5 years were underweight in 2009, 48% were stunted, and 20% had died due to nonavailability of a proper diet (Table 3.2). These facts reveal that accessing food alone does not guarantee food security, and to be food-secure, people must have safe and healthy food to meet their physical requirements and lead a healthy and productive life.

3.2 The Horticulture Sector and Its Implications for Developing Countries

The horticulture sector has become one of the major economical hubs in developing countries, including the Asia-Pacific region. This sector deals with the production of fruit, vegetables, flowers, and medicinal plants as an emerging economic activity of the region. Fruit and vegetables, which are part of the daily diet, are grown in more than 50% of the world's acreage in this region. Diversified production of fruit and vegetables provides rich sources of minerals, vitamins, and micronutrients like zinc, copper, and manganese, which help to regulate different biological functions for

Table 3.1 Estimated growth in world's population and food consumption.

Countries	Average annual growth rates (%) 1970–2000			Average annual growth rates (%) 2000–2030			Average annual growth rates (%) 2030–2050		
	kcal/person	Population	Food consumption	kcal/person	Population	Food consumption	kcal/person	Population	Food consumption
World	0.49	1.70	2.20	0.29	1.03	1.32	0.15	0.48	0.63
Developing countries	0.77	2.05	2.83	0.36	1.20	1.56	0.18	0.57	0.75
Sub-Saharan Africa	0.15	2.80	2.95	0.57	2.23	2.81	0.42	1.48	1.91
North Africa	0.00	2.57	2.57	0.17	1.56	1.74	0.09	0.82	0.92
Latin America and Caribbean	0.74	2.02	2.77	0.32	0.94	1.26	0.13	0.28	0.40
South Asia	0.47	2.23	2.71	0.51	1.29	1.81	0.33	0.53	0.86
East Asia	0.49	1.48	1.97	0.35	0.47	0.82	0.06	−0.17	−0.10
Industrial countries	1.19	0.74	1.94	0.07	0.47	0.54	0.03	0.13	0.16
Transition countries	0.41	0.80	0.49	0.28	−0.64	−0.37	0.19	−0.78	−0.59

kcal = kilocalorie.
Source: Moir and Morris (2011).

Table 3.2 Malnutrition of children in South Asia (%).

Countries	Underweight	Stunted growth	Wasted/dead
Afghanistan	32	60	8
Bangladesh	40	43	18
India	44	48	20
Nepal	39	50	12
Pakistan	30	49	14
Sri Lanka	20	17	13

Source: Adapted from FAO (2012).

human health. This nutritive profile of fruit and vegetables may vary in accordance with their variety, type of agronomic practices, and harvesting methods.

The rise in income and in living standards has accelerated demand in this sector in most of the developing countries. But lack of proper produce-handling methods, poor infrastructure for storage and processing, results in waste of the produce between 10% and 40%. This restricts movement of the produce at the domestic level and for export purposes. The concept of post-harvest loss (PHL) was first proposed at the World Conference held in Rome during 1974, which later in 1980 considered the exposure of fruit and vegetables to PHLs to be of prime importance. Clearly, large-scale production of fruits and vegetables does not have much value unless it is consumed by the end user. But from production to consumption, the supply chain of fresh produce follows several channels that cause losses both in quality and in quantity, rendering it unacceptable for consumption.

3.3 The Horticulture Sector in Pakistan

Horticulture is an important subsector of Pakistan's agricultural economy. According to the Ministry of National Food Security and Research (MNFS&R) the total area under horticultural crops is currently about 1.32 million hectares, with fruits and vegetables' share being almost 58% and 42%, respectively, during 2014–2015. The country produces about 10.41 million tons of fruit and vegetables, of which fruit contributes 7.01 million tons (50%) and vegetables including condiments 7.11 million tons (50%). Horticultural crops contributed about 12% to the national agricultural GDP during the last decade.

Analysis by province of fruit area and production (Figure 3.1) shows that Punjab province leads in both areas (47.2%) and production (63.7%). Regarding

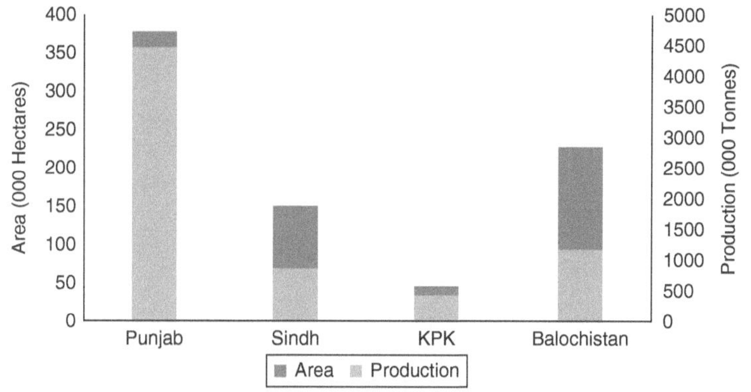

Figure 3.1 Area and production of fruit, by province (2014–2015) (GoP 2023).

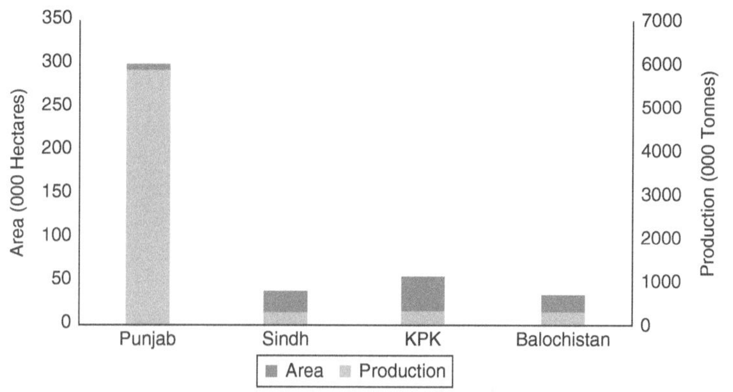

Figure 3.2 Area and production of vegetables, by province (2014–2015) (GoP 2023).

the distribution of vegetable area and production (Figure 3.2), Punjab remains the major contributor, with 67.6% in area and 81.3% of production.

The main fruits include citrus, mangoes, dates, guava, banana, and apples. However, Pakistan also produces good quality pears, peaches, plums, apricots, and grapes. Among the important vegetables and spices are potatoes, onions, tomatoes, chilies, garlic, and a range of leafy and other vegetable crops. The distribution of percent area and production for major fruits and vegetables is given in Figures 3.3–3.6.

In past, Pakistan exported fruit and vegetables for about 376.5 million dollars and 159.7 million dollars, respectively, during 2014–2015. On a comparative basis, fruit trees are perennials and involve less management and labor compared to vegetables, which are more intensive and require higher labor and inputs.

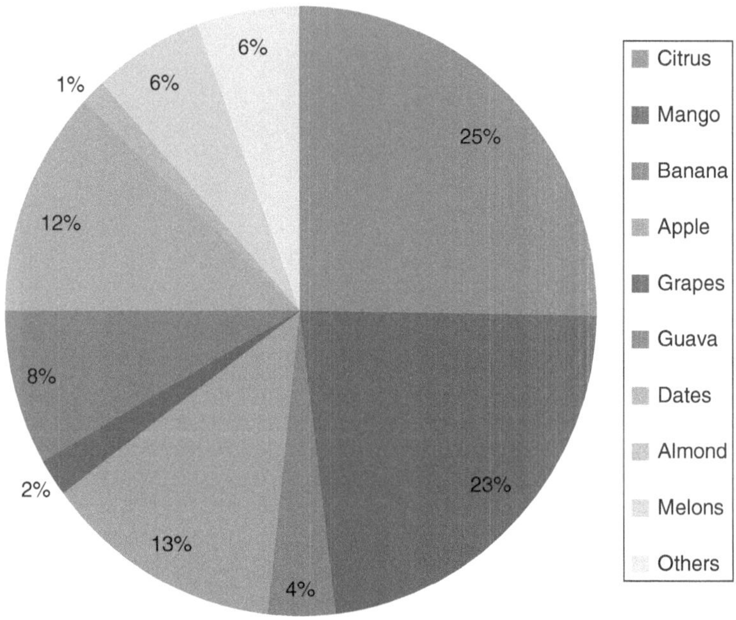

Figure 3.3 % share of fruits under total cultivated area (2014–2015) (GoP 2023).

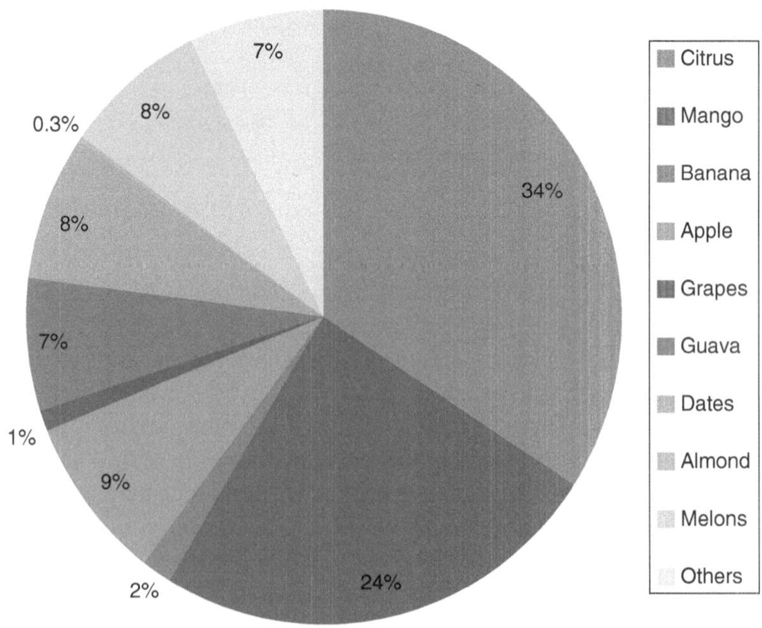

Figure 3.4 % share of fruits in total production (2014–2015) (GoP 2023).

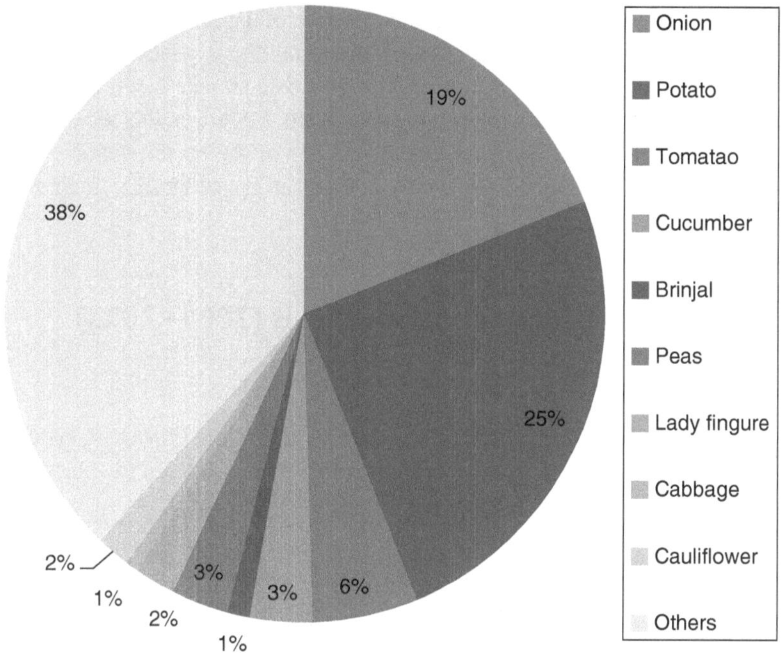

Figure 3.5 % share of vegetables under total cultivated area (2014–2015) (GoP 2023).

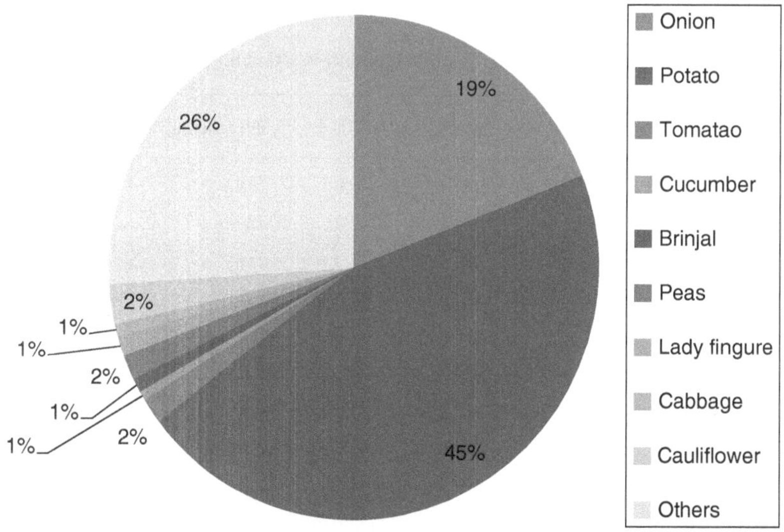

Figure 3.6 % share of vegetables in total production (2014–2015) (GoP 2023).

Horticulture produce has tremendous potential for export to several countries, but intensive PHLs limit this potential due to a lack of packing/grading material, inappropriate transport and cold storage facilities, and a significant portion of the fresh produce is wasted. Moreover, this sector provides employment by requiring intensive labor in the production system, thereby increasing individuals' economic status by generating high net returns per unit area.

3.4 Fresh Produce Industry Analysis (2005–2015)

3.4.1 Area and Production (Fruits)

Over decades, the fruit industry has been a main strength of Pakistan. Provincial trends in total area and production of fruit are given in Tables 3.3 and 3.4.

3.4.2 Area and Production (Vegetables)

The production of vegetables is extensive in each of the four provinces, but significantly more so in the Punjab. Although a wide range of vegetables is grown, the most important include potatoes, onions, chilies, tomatoes, turnips, lady fingers, and peas. The trends in total area and production of vegetables are presented in Tables 3.5 and 3.6.

Table 3.3 Provincial area under fruit production in Pakistan *('000 Hectares)*.

Year	Punjab	Sindh	KPK	Balochistan	Pakistan
2005–2006	386.7	135.4	46.8	245.6	814.5
2006–2007	395.2	143.4	46.9	247.4	832.9
2007–2008	400.9	150.1	47.2	255.2	853.4
2008–2009	400.1	154.9	47.3	254.7	857.0
2009–2010	395.9	157.8	47.0	251.8	852.5
2010–2011	390.0	154.7	43.3	248.0	836.0
2011–2012	387.6	142.4	43.1	249.4	822.5
2012–2013	385.7	146.0	43.5	235.9	811.1
2013–2014	380.2	149.4	44.8	237.9	812.3
2014–2015	378.4	149.4	45.6	228.2	801.6

Source: Adapted from MNFS&R, Agricultural Statistics of Pakistan (GoP 2023).

Table 3.4 Provincial fruit production in Pakistan *('000Tonnes/Hectares).*

Year	Punjab	Sindh	KPK	Balochistan	Pakistan
2005–2006	4764.3	884.3	519.1	980.2	7147.9
2006–2007	3744.9	888.8	516.1	861.4	6011.2
2007–2008	4556.1	984.1	516.4	1122.0	7178.6
2008–2009	4337.9	1015.4	522.4	1175.7	7051.4
2009–2010	4494.9	1007.1	447.2	992.1	6941.3
2010–2011	4374.6	1017.2	412.9	1121.8	6926.5
2011–2012	4263.5	932.6	414.4	1204.5	6815
2012–2013	4083.3	930.7	415.6	1133.9	6563.5
2013–2014	4191.4	944.1	419.5	1205.2	6760.2
2014–2015	4473.9	942.9	424.6	1181.6	7023.0

Source: Adapted from MNFS&R, Agricultural Statistics of Pakistan (GoP 2023).

Table 3.5 Provincial area under vegetable production in Pakistan *('000 Hectares).*

Year	Punjab	Sindh	KPK	Balochistan	Pakistan
2005–2006	240.4	37.1	47.5	38.6	363.6
2006–2007	256.8	34.5	48.9	38.8	379.0
2007–2008	279.8	35.7	48.7	44.1	408.3
2008–2009	269.3	36.9	49.3	42.2	397.7
2009–2010	263.5	36.7	47.6	40.4	388.2
2010–2011	285.2	38.7	45.6	41.5	411
2011–2012	310.8	41.2	46.3	41.5	439.8
2012–2013	301.3	47.3	47.8	37.1	433.5
2013–2014	288.6	51.8	51.1	36.9	428.4
2014–2015	299.5	52.3	56.0	35.0	442.8

Source: Adapted from MNFS&R, Agricultural Statistics of Pakistan (GoP 2023).

Table 3.6 Provincial vegetable production in Pakistan *('000 Tonnes/Hectares).*

Year	Punjab	Sindh	KPK	Balochistan	Pakistan
2005–2006	3475.5	211.4	520.1	485.7	4692.7
2006–2007	4483.7	215.1	532.0	488.7	5719.5
2007–2008	4442.1	238.3	527.6	467.8	5675.8
2008–2009	4842.4	249.7	533.5	529.6	6155.2

(*Continued*)

Table 3.6 (Continued)

Year	Punjab	Sindh	KPK	Balochistan	Pakistan
2009–2010	5007.5	249.5	492.7	436.5	6186.2
2010–2011	5338.3	265.8	479.3	520.5	6603.9
2011–2012	5266.9	269.7	501.2	528.9	6566.7
2012–2013	5616.0	326.7	506.4	448.6	6897.7
2013–2014	4733.1	350.6	518.02	445.9	6047.62
2014–2015	5792.9	356.9	558.3	418.7	7126.8

Source: Adapted from MNFS&R, Agricultural Statistics of Pakistan (GoP 2023).

3.5 Constraints on Production and Postharvest Management

Horticultural crops are of major importance in some areas of Pakistan, while they remain neglected in other parts, where people grow these crops only to fulfill local fruit and vegetable requirements rather than producing for export markets; hence, crop yields but not quality remain a priority. Furthermore, a poor postharvest handling and marketing system leads to lower net profits in the overall supply chain, while consumer prices both in local and international markets are rising. The local horticulture industry is currently facing a number of challenges, among which three main aspects are very important: (i) large yield gap: in comparison with other countries (Table 3.7), (ii) high postharvest losses, and (iii) export price gap.

3.6 Postharvest Losses

PHLs refer to the measurable loss in food quantity and quality from harvest to consumption by the end user. Losses in food quality refer to the reduction in its nutrient contents and acceptability by the consumer. These are more common in developed countries (Kader 2002). Losses in quantity refer to the reduction in the amount of the product. These are more common in developing countries due to poor sanitation and handling facilities (Kitinoja and Gorny 1999). It is estimated that about 12 MT of fruits and 21 MT of vegetables are lost every year, corresponding to a loss of 240 billion rupees. Interventions in reducing PHL are therefore necessary to address the key

Table 3.7 Yield gap and ranking of important horticultural commodities.

Crop	Max. yield in the world (Tons/ha)	Yield of Pakistan (000 Tons/ha) and ranking	Yield gap (Tons/ha)
Citrus	47.0 (USA)	9.03 (19th)	37.97
Mango[a]	26.7 (Guatemala)	10.7 (13th)	16.0
Date	33.77 (Egypt)	7.61 (4th)	26.16
Apple	69.85 (Slovenia)	6.94 (35th)	62.91
Apricot	24.35 (Slovenia)	7.46 (18th)	16.89
Grapes	23.31 (Israel)	3.77 (34th)	19.54
Peach	22.42 (Slovenia)	7.4 (26th)	15.02
Pear	47.71 (Slovenia)	13.26 (20th)	34.45
Plum	25.91 (Czech Republic)	8.2 (26th)	17.71
Potato	48.39 (Belgium)	17.64 (32nd)	30.75
Onion	55.07 (USA)	13.3 (35th)	41.77

[a] For countries with an area over 1000 ha.
Source: Adapted from FAO (2004).

challenge of food security. Postharvest and value addition are considered as an integral part of all strategies developed to combat food insecurity of the growing population.

3.7 Factors Responsible for Postharvest Losses

PHLs may occur due to internal and external factors which are described as follows.

3.7.1 Internal Factors

- Harvesting of the crop usually requires taking many parameters into consideration, like determining the maturity index, weather conditions, harvesting operations, handling, and storage of the produce for further utilization in the supply chain system. Selection of the appropriate time for crop harvesting determines the degree of initial loss to the produce. Harvesting time relies on an already developed maturity index: any fluctuation or absence of available maturity index may cause a loss to the crop. Environmental conditions including rainfall can also severely affect harvesting time and its duration by interrupting its regular operations.

These may result in addition or reduction of moisture content in the harvested produce, deteriorating its quality. Handling and harvesting operations also play a crucial rule in this aspect because poor handling or lack of proper tools and storage conditions required after harvest can reduce quantity and quality of the crop during harvest.

- Pre-cooling is one of the most important operations that are applied to fruit and vegetables after harvest. Any variation in the timing of pre-cooling and any lack of information regarding its application to the produce may result in loss of the produce before going to the end user. In Pakistan, the high upfront cost and seasonal usage of pre-cooling technology do not justify the investments in this sector. Although rental services are available, expensive payment structure makes it unfeasible for most producers.
- Proper grading and packing of the produce are critical to minimize the mechanical injury to the produce. Proper packing during transportation from rural to urban areas and then for exporting purposes are necessary to ensure good quality of the produce. In developing countries, which suffer from poor packaging and grading systems, most of the fruit and vegetables undergo PHLs during this stage. Generally, harvested produce is packed in the field without any pretreatment, and the most common factor responsible for losses here is the poor quality and inappropriate size of the packing container. Containers made of wooden material or sharp plastic inflict deep cuts to fruit and vegetables which result in water loss and rapid decay of the produce. In the case of vegetables, growers transport their produce from the field to market even without packing, which results in wastage at different stages. In Pakistan, fruits are packed in low-grade wooden containers which may cause damage and puncture to the product, and subsequent delays in transportation may result in quality loss of the product.
- Transportation has always been of main concern for growers to send their harvested produce to the market within appropriate time to get its good price. In developing countries, poor roads infrastructure and small holdings make it tough for growers to transport their horticultural crops according to their shelf life. Furthermore, the majority of growers in developing countries do not have their own vehicle and depend on other growers and associations to transport their produce to market, which ultimately cause deterioration of fruit and vegetables (Kader 2002). Sometimes, to reduce transportation expenses, growers pile containers up on each other during transportation, and this results in compression of produce which causes bursting or crushing. In Pakistan, the transportation of horticultural commodities is done in open trucks. During long travels,

the product may be subjected to high temperature which may result in losing product freshness and spoilage. As per estimate, 10% of produce damages during a conventional method of transportation. The transportation through refrigerated trucks is limited as the cost is significantly higher as compared to open body trucks/pickups.

- Storage facilities, including cleanliness, storage structures, monitoring of temperature and humidity, regulation of pest populations, and prevention from diseases determine how much of the stored produce is lost. Poor management of all these factors can lead to deterioration of the storage structure as well as of the quality of the produce. Although limited cool storage facilities are available in Pakistan, it is important to identify and understand the bottlenecks of this domain.
- Fruit and vegetables have a specific shelf life, during which their nutrient contents and other mineral profiles are naturally maintained. However, several factors like temperature, relative humidity, wind flow, and sanitation facilities during transportation and storage can greatly influence the nutritive value of the produce. And poor handling may cause development of microorganisms inside the produce that can directly affect the quality of the food, making it unacceptable to the consumer.

3.7.2 External Factors

- Exposing harvested produce to external environmental factors can result in deterioration of fresh produce. Keeping produce under direct sunlight or exposing it to the cold for too long can cause over-ripening or softness of fruit that makes it unacceptable to the consumer. In developing countries, due to lack of resources and unavailability of packaging units, growers tend to use the same packing or transportation containers for multiple produce. This practice is extremely damaging to the fruit's physical and chemical properties. Using the same containers may result in flavor contamination; and inappropriate container size can cause surface damage and in many cases discoloration of the fruit.
- The most obvious factors responsible for PHLs include attacks by insect and rodent pests. These pests invade the packaging material and get transported by carriers to other places, infesting fruit, and vegetables to a great extent. They sometimes cause heavy losses to produce during transportation. After an initial damage, the produce becomes vulnerable to disease infestation. Disease occurrence mostly results in decay and rotting of the produce, which also contaminates fresh produce during transportation and marketing.

3.8 Strategies to Reduce PHLs in Fruits and Vegetables

There are different strategies for different stages which are explained in Table 3.8:

Table 3.8 Strategies to reduce PHLs in fruits and vegetables.

Stages	Actions needed
Harvesting	During harvesting, mechanical injuries may occur in produce. Careful harvesting should be done as much as possible to minimize scratches, bruises, and punctures in fruits and vegetables. Early morning is the most appropriate time for harvesting, if later in the day during hot span harvesting is needed, produce should be kept in shade after harvesting.
Handling	Handling must be given full attention in the case of fruits and vegetables because their skin is the most suitable barrier between microorganisms and many insect pests to invade the crop. Poor handling may cause mechanical injury to the skin of the fruit and may help the establishment of pests and disease to the produce.
Grading and cleaning	Poor grading of the produce can greatly reduce its shelf life, freshness, and quality. However, appropriate grading and cleaning of the good produce from the bad ones can reduce losses and help in reducing marketing cost due to proper operations suitable for transportation and marketing of crop.
Packaging	Packaging of the crop is the most important operation that starts right from the crop harvesting to the consumption by the end user. Packaging containers and its material determines the freshness and deterioration of fruits and vegetables during field operations, transportation, and marketing and storage operations. Suitable packing material in the case of leafy vegetables and fruits should be used to maintain its quality for marketing purposes.
Transportation	Transportation is the most vital component of a supply chain system and includes movement of the produce from farmer's field to market or storage facilities. During transportation, special attention should be given to the selection of the vehicle and loading and unloading facilities. Produce must be transported during the cooler part of the day and must not be watered before loading and transporting. Prior watering may result in decay and deterioration of the produce.

Table 3.8 (Continued)

Stages	Actions needed
Storage	Storage is the most common practice adopted by growers all over the world. It should be noted that only crops of good quality should be stored by maintaining their proper temperature, relative humidity, and other factors necessary for its storage. Storage of damaged crops would result in further deterioration of the crop and may affect healthy food also.
Processing	Increase in financial matters of the people has diverted their attention to consuming processed food to meet their daily intake. Processing is a step where value-added activities are being carried out in each food to enhance its digestibility and stability for regular use. Appropriate processing facilities should be used to maintain the shelf life of the food.

Source: Adapted from Kiaya (2014).

3.9 Recent Advances in Postharvest Technology

In recent years, much attention has been given to finding out ways that can limit the extent of PHLs imposed on fresh produce. The growing attractiveness of processed food, due to its rapid digestion and nutritive value, has triggered the conversion of fresh produce into processed food after going through different technological processes. This processing not only reduces produce losses but also increases its shelf life and dietary range. In order to reduce PHL, the following points need special attention.

3.9.1 Breeding and Provision of Raw Material

With advances in biological science, including genetics and biotechnology to manipulate living matter for the benefit of human beings, crops with highly attractive attributes have been developed by scientists. This has not only triggered cultivation of crops with high production potential but has also developed resistance against several insects and diseases that can cause losses to produce. Provision of suitable and good quality raw material, including seeds and root stock, is considered an important prerequisite for achieving high production with minimal losses. Selecting good varieties and performing appropriate agronomic practices help realize a crop's high production potential. Close to harvesting, the maturity index needs to be followed and proper harvesting methods should be adopted to reduce PHLs.

3.9.2 Cold Supply Chain System

This system involves storage of fresh produce at controlled temperature for a sufficiently long period that can maintain its quality and can yield more profit if sold during a period of produce shortage. This technology is more common in developed countries, where growers have cold storage facilities near their farms, and where they can store their fruit and vegetables, eliminating the risk of losses over time. These cold storage structures also link growers with end users through market information about the demand for the produce: this establishes a chain reaction in storage and selling of the fresh produce. These cold storage structures are very expensive and need continuous power supply to maintain a low temperature. But low-income levels of growers and power cuts in developing countries make them difficult to operate and give satisfactory results. However, some cold storage structures can be observed in areas controlled by a farming cooperative organization operating on behalf of all the growers associated with it in that locality, and where growers pay to store their produce in its cold storage structure.

3.9.3 Processing Units

A range of valuable fruits and vegetables are produced in developing countries that are unique in taste and color. Most fresh produce has a medicinal and nutritive value. As people are becoming more conscious about their health and daily dietary needs and tend to avoid chemically processed foods, the processing of indigenous fruit and vegetables as per required entity can fulfill the nutrition and dietary needs of the people and can give better consumption of the produce. Sometimes, direct consumption of the produce is not appreciated, while products associated with its processing can receive more interest from the consumer. Installing processing units within a country can thus allow more consumption of the produce in the form of different processed products and can also limit the dependency on imported processed foods.

3.9.4 Development of Marketing Strategies to Address PHLs

Marketing strategies are developed with the aim of selling the fresh produce in time to achieve maximum financial returns. But this practice is more complex with fruit and vegetables compared to other goods because these are soft and perishable. These factors limit free flow of transportation to a greater distance as well as time limit since its harvesting. The following factors should be considered before developing marketing strategies in accordance with the local environment.

- **Variation in production patterns**
  The production volume of any produce depends on market forces that determine the future cultivation volume of produce. The rise in prices during low supply periods encourages growers to grow a greater area that supplies a larger volume of produce to the market during the following season but offers a lower price to growers due to surplus availability. This trend is mostly seen with vegetable production, while fruit cultivation is less susceptible to this phenomenon. In developing countries, low education levels and growers' lack of awareness regarding the market system make it more complex to follow such swings in supply and demand and play the system with production strategies. But educating farmers about market needs and dynamics can result in variation among production volumes of the produce. Using the information on price variations in different parts of the country can help farmers to understand the balance between demand and supply and aid to plan to transport their produce to alternative localities for better returns.
- **Market information**
  Accurate and adequate market information is the key component in getting good prices and developing strategies for the efficient utilization of resources. It also helps to circulate produce among different regions within countries and for exports as well. Market information includes produce demand, the volume available in the market, price variations with respect to season, and future demand of the produce. Based on these, growers can design their production system and can decide the time of harvest and its transporting facilities in response to market demand. Market information also helps transporters and storage operators to engage their staff and resources to react as per situation demand accordingly. It also provides a baseline for government agencies to develop a comprehensive agricultural market policy for food and for assessing its economic development and export needs.
- **Developing a strategy for the market**
  There are several methods available for developing a market strategy for a given product. But the successful marketing strategy is based on a marketing situation specific to a locality for which it is being addressed. For developing a strong and reliable market strategy, the following steps should be followed.
- **Baseline survey of the existing marketing situation**
  The availability of the current market situation is necessary to start with market situation analysis. A baseline survey must be conducted to have

an idea about the strengths and weaknesses of the marketing system. These surveys are mostly conducted by government-chosen agencies through a series of direct approaches to market associates. They extract complete data on the low pace of market growth, inappropriate service facilities, poor organizational setup, lack of interaction among growers and demand for the produce, association of international trading policy with the local one, and all other factors which are responsible for poor local marketing. The collected data are analyzed by relevant experts, keeping in view the local and world marketing trends of the produce.

- **Planning for the future**
  Based on the data gathered, the weakness and lack of interconnections in the organizational setup, a way forward is suggested supporting local conditions. The gaps are sorted and organized based on their importance; goals should be set to be achieved, and organizational authority should be designated as a monitoring factor. Comprehensive efforts were made in the past for the sectorial SWOT analysis highlighted that weakness in the regulatory framework constrains the sector's productivity and profitability. Also, lack of comprehensive horticultural policy indicating board direction, institutional profiling, and coordination for innovation has been a major constraint for the sector. Sometimes, it becomes difficult for government agencies to get directly involved in this system for follow-up practices, so any responsible agency with good knowledge of marketing fruits and vegetables should be hired as a follow-up authority. This would get information and assess the strength of developed and implemented plans for the marketing system.
- **Increasing access to International Market**
  Outmoded production technology and marketing methodology along with poor agglomeration and lack of traceability have made it difficult for developing countries to meet the stringent sanitary and phytosanitary (SPS) requirements of importing countries. With existing infrastructure, it is difficult to develop reliability in the international market as noncompliance with standards may result in a temporary or permanent ban from importing countries. Also, reliance on pesticides for pest management and lack of information regarding environmental conservation through use of safer plant protection products imposesconcerns related to maximum residue limits (MRLs). Furthermore, limited information on integrated pest management (IPM), integrated crop management (ICM), rational use of pesticides, and harvest intervals have contributed to noncompliance of good agricultural practices (GAP) in the horticulture sector. Satisfaction to SPS measures along with capacity-building support and supply chain management infrastructure can provide opportunities to capture the international market for counter-season export with an edge over the competition.

- **Training of market personnel**
  Lack of professionals has always been a main concern for developing countries to achieve any goal. This inhibits implementation of improvements in the marketing system. All the personnel engaged in developing and implementing marketing improvements must be trained in all respects to evaluate the situation on a sound basis. Training of all market entities including growers, transporting, and packaging agencies, organizational resources of the markets and associates engaged in produce handling must be conducted in a systematic way for successful implementation of the strategy. This can be done with the help of multilateral and bilateral international programs which involve international experts from developed countries aimed to build the institutional and scientific capacities of stakeholders for modernizing sustainable production systems.

References

FAO (2002). *The State of Food Insecurity 2001*. Rome: Food and Agriculture Organization.

FAO (2004). *FAO Statistical Yearbook 2004*. Rome: Food and Agriculture Organization.

FAO (2006). *World Agriculture: Towards 2030/2050. Interim Report*. Rome: Food and Agriculture Organization.

FAO (2012). *FAO Statistical Yearbook 2012*. Rome: Food and Agriculture Organization.

GoP (2023). Agriculture statistics of Pakistan, Chapter No 2. https://www.pbs.gov.pk/agriculture-statistics-tables (accessed 13 November 2023).

Kader, A.A. (2002). *Postharvest Technology of Horticultural Crops*, 3e. Oakland: University of California Agriculture and Natural Resources. Publ. 3311.

Kiaya, V. (2014). Post harvest losses and strategies to reduce them. Action Contre la Faim (ACF), member of ACF International, p. 25.

Kitinoja, L. and Gorny, J.R. (1999). *Postharvest Technology for Small-scale ProduceMarketers: Economic Opportunities, Quality and Food Safety*. University of California Postharvest Horticulture. Series No. 21.

Moir, B. and Morris, P. (2011). Global food security: facts, issues and implications. *ABARES Science and Economic Insights* (1): 1.

United Nations (UN) (2010). *The 2010 Revision of World Population Prospects*. New York: UN.

World Bank (2009). *Guidance for Responses from the Human Development Sectors to Rising Food Prices*. Washington, DC: World Bank.

4

Climate Change and Agricultural Productivity

Ghaffar Ali

College of Management, Shenzhen University, Shenzhen, China

4.1 Impacts of Climate Change on Food Security in Developing Countries

Climate change is defined by the Intergovernmental Panel on Climate Change (IPCC) as "any change in climate over time, whether due to natural variability or as a result of human activity" (Shahzad and Amjad 2022). This term, climate change, can also be mentioned or stated as "a change in the statistical distribution of weather patterns when that change lasts for an extension time" (Orgeret et al. 2022). Sometimes, it is referred to a long-term change in the weather conditions and/or fluctuations in average weather conditions. Global change showed noticeable impacts, almost in the past 50 years, related to agriculture in developing countries. Over 1 billion people are supported in their livelihoods by the present food system, which includes production, transport, processing, packaging, storage, sale, consumption, loss, and waste (Din et al. 2022). The amount of food available per person has grown by more than 30% since 1961, along with increasing usage of nitrogen fertilizers (up nearly 800%) and water resources for irrigation (up more than 100%) (Boretti and Rosa 2019). However, 613 million women and girls between the ages of 15 and 49 years are iron-deficient, 821 million people are thought to be undernourished, 151 million children under the age of 5 years are stunted, and 2 billion adults are either overweight or obese (Petry et al. 2019). The food system is under stress from both climate change and nonclimate stressors, such as population and

Food Security in the Developing World, First Edition. Edited by Muhammad Khalid Bashir, Steven G.M. Schilizzi, and Ghaffar Ali.

wealth development and the demand for items derived from animals. The food system is responsible for between 21% and 37% of all greenhouse gas (GHG) emissions (Crippa et al.2021). These are related to agriculture, including land productivity, and to transportation, storage, packing, processing, and consumer goods (medium confidence). This estimate comprises emissions of 9–14% from field gate livestock and crop production operations and 5–14% from land use and land-use change, involving deforestation and peatland degradation (high confidence); 5–10% comes from supply chain operations (medium confidence) (Xu et al. 2021). GHG emissions from food loss and waste are included in this calculation.

Agriculture produce the majority of the energy consumption from the supply side of the food supply chain between 2007 and 2016, with crop and livestock activities at the farm gate producing, including both, 142 Tg methane (CH_4 yr^{-1}) (high confidence) and 8.0 Tg nitrogen dioxide (N_2O yr^{-1}) (high confidence), and carbon dioxide (CO_2) emissions linked to applicable land-use change dynamics, like deforestation and peatland degradation, producing 4.9 Tg CO_2yr^{-1}. According to the IPCC AR5's 100-year GWP values (climate feedback), total GHG emissions from agriculture were 6.2 1.4 $GtCO_2$-eq yr^{-1}, rising to 11.1 2.9 $GtCO_2$-eq yr^{-1} when relevant land use was taken into account (Hong et al. 2021). Without any action, they are probably going to rise by 30–40% by 2050 as a result of rising demand brought on by changes in eating habits, population growth, and income levels (Nugee 2022) (high confidence).

Agriculture and the food system are essential to global climate change solutions (Uhlenbrook et al. 2022). Reducing GHG emissions and strengthening the resilience of the food system may be achieved by combining supply-side measures like efficient production, transportation, and processing with demand-side initiatives like changing dietary preferences and reducing food loss and waste (Lu et al. 2022). Large-scale land-based adaptation and mitigation methods may be implemented with the help of these integrated approaches, all without endangering food security due to heightened competition for agricultural land and rising food costs. Negative consequences would include a rise in the number of malnourished individuals and negative effects on smallholder farmers if food system reforms in farm management, supply chains, and demand were not coupled (medium evidence and high agreement). All that is required to remedy these impacts are transitions.

Greenpeace Thailand revealed some effects of climate change that are occurring, namely sea levels are rising due to melting glaciers and global temperatures are rising, and these are more vulnerable to extreme weather events such as heat waves, droughts, and floods, for example. During the last two and half decades, climate of Pakistan is changing, and it has drastic effects on all the sectors of agriculture in different ways (Rehamn 2015). The impacts will knock out the low-income countries critically, such as developing countries of Africa and Asia-Pacific that have the limitation of protecting themselves resulted from sea level rise, spread of germs, and lower agricultural output[1]. Table 4.1 presents the literature summary of climate change impacts (ecology, economy, and health- wise comparison) in Africa and Asia-Pacific.

Likewise, food production and food prices are directly affected by climate change and ultimately turn into problem of food security. World population projection shows that until 2080, the population of the world will reach 9.2 billion from 5.9 billion, so demand for food is predicted to rise by about 300% mainly due to the larger population by 2080 (Ali et al. 2021). Hence, this rise is likely to create an inversely proportional relationship between demand and supply of food. If this is the case, a decline in food production will increase pressure on food prices.

IPCC claimed that an increase in temperature by 2°C beyond preindustrial levels leads to a fatal risk of severe climate change-related impacts and a spike in current rate of global temperature from 0.2 to 0.3°C (Hoegh-Guldberg et al. 2018). The current level of carbon dioxide (CO_2) concentration is 430 parts per million (ppm) which is increasing almost by 2 ppm annually.

Climate change impacts can be divided into three categories, i.e. ecology, economy, and health. Table 4.1 shows such a comparison based on Supaporn's categorization. Generally, in Africa, the impacts of climate change are increasing in three dimensions, i.e. extreme weather events, deterioration of food security, and risk of health and life. Meanwhile, few impacts are also obvious in Asia: the decreasing resources of water, food security, agriculture vulnerability, and increasing diarrheal disease. These are the examples of the impacts of climate change in the latest scenarios. Furthermore, comparison showed that most of the current statuses in each criterion are almost similar in Africa and Asia. Considering the ecology of climate change impacts, temperature is higher and drier in many areas; an increase in water stress and deforestation have an effect on the ecosystems of both regions. Likewise, considering the economy, there are more risks in agriculture production and food security and decreasing trend in crop yields that may lead to the risk of food hunger. There are projections of the development constraints in Africa and projections of economic losses in Asia (Bashir and Schilizzi 2013).

1 A UN's commission designated especially abbreviated in full as: United Nations Framework Convention on Climate Change.

Table 4.1 Climate change impacts in Africa and Asia-Pacific (literature summary).

Impacts	Africa	Asia	Sources
Ecology			
Temperature and precipitation	• Higher warming and drier that caused flood and drought in some areas. • Increasing variant weather such as rainfall, and annual rainfall decrement in the Mediterranean Africa, Southern Africa, and North side of Sahara while increase in East Africa.	• Higher warming in central, northern, eastern, and southern Asia. • Increasing life-threatening conditions such as floods, droughts, forest fires, and tropical cyclones. • More glacier and snow melting in some areas such as Himalayas.	Ansari and Grossi (2022), (Crippa et al. (2021), Boretti and Rosa (2019)
Water	• Increasing water stress for many countries. • Prediction of increasing severe water shortages by 2020.	• Increasing water stress and decreasing freshwater availability. • Agriculture productivity decrement along with aquaculture mainly because of water stress.	
Forest and ecosystem	• Increasing the degradation of grasslands, deforestation, and forest fires.	• Increasing deforestation that cause more risk of extinction for many species and habitat fragmentation.	
Economy			
Agriculture and food security	• Decreasing grain yields and diminishing food security that affect the increasing people's quantity which is at risk mainly due to hunger problem.	• Increasing land degradation and desertification that affect agricultural production. • Food security is also vulnerable, and chances of	Meyers and Kalaitzandonakes (2015), Leakey (2018), Nhamo et al. (2019), Stuch et al. (2021), Noort et al. (2022)

Table 4.1 (Continued)

Impacts	Africa	Asia	Sources
	• Loss of land and shorter growing seasons' effect to decreasing agricultural production and importing production in Africa.	lessen are high in several countries due to drought situations embedded with floods, rise in sea level, and many more.	
Economy projection	• Projection of the constrain development in Africa caused by increasing droughts, floods, and extreme events that affect stresses on water resources, food security, human quality of life, and infrastructure systems.	• Projection of extreme event will be 66–90% chance of rain and potentially severe storms in the next decade that affect the nation economy, for example, 10 years ago in Thailand, there is the long summer drought that caused economic losses of total more than 70,000 million baht.	
Health			
Health	• Increasing the expansion of a host of contagious ailments, for instance, dengue fever, malaria, meningitis, and cholera that affect human health (and now Africa is experiencing the effects of AIDS already)	• Increasing exposure to vector-borne infectious diseases and heat stress can affect human health • Increasing diarrheal disease in South and Southeast Asia, and toxicity of cholera in South Asia.	Kohnert (2021), Gostin (2021), Sabin et al. (2020), World Health Organization (2018)

According to the health impacts analysis, because of the heat and temperature changes, there are more health risks from diseases due to the increasing expansion of a host of infectious diseases.

Food security is directly affected by climate change, and the necessity to mitigate it may raise competition for resources used in agriculture. By potentially increasing competition for land, addressing climate change through the deployment of land-based technology for negative emissions based on biomass production would put growing pressure on food production and food security (Figure 4.1). Figures 4.2–4.4 depict the scenarios of climate change impacts, food, agriculture, and adaptation to climate change in Africa and Asia-Pacific regions. It is evident that agriculture productivity in the study regions is seriously vulnerable to climate change regarding agricultural productivity in the next 50–60 years, especially in South Asian, Southeast Asian, and South African regions of the world. Increase in prices of different crops, namely wheat, maize, rice, and other crops, will be 75%, 89%, 90%, and 83%, respectively, by 2030 (Figure 4.3). Also, the trend in growth rate will be decreasing downward by the end of 2030. Moreover, average yields, farming activities, and agricultural productivity are significantly vulnerable in lower parts of Africa and Asia-Pacific, especially Afghanistan, Australia, Bangladesh, India, Lao PDR, Pakistan, and Thailand (Figure 4.4) by multiple means in the next few decades. The following parts of this chapter are comprised of multiple-case studies drawn for different vulnerable developing countries to get more insights on climate change and agriculture allowing multiple scenarios.

This chapter discusses how climate change impacts agricultural food security, the alternatives available to the food system to adapt and mitigate, the synergies as well as trade-offs among all these options, and the prerequisites for their adoption. It really does, so use a food system perspective. The possibility of integrating supply-side policies like sustainable intensification (raising production per hectare) and food security policies to contribute to climate change mitigation is evaluated in this chapter, along with the importance of incremental and transformative adaptation.

4.2 Case Studies

This section comprehends the case studies from different Asian countries such as China, India, Myanmar, Pakistan, and Thailand. Altogether, case studies focus on climate change vulnerability to agriculture, and few also consider food security, natural environment, human settlements, and social vulnerability.

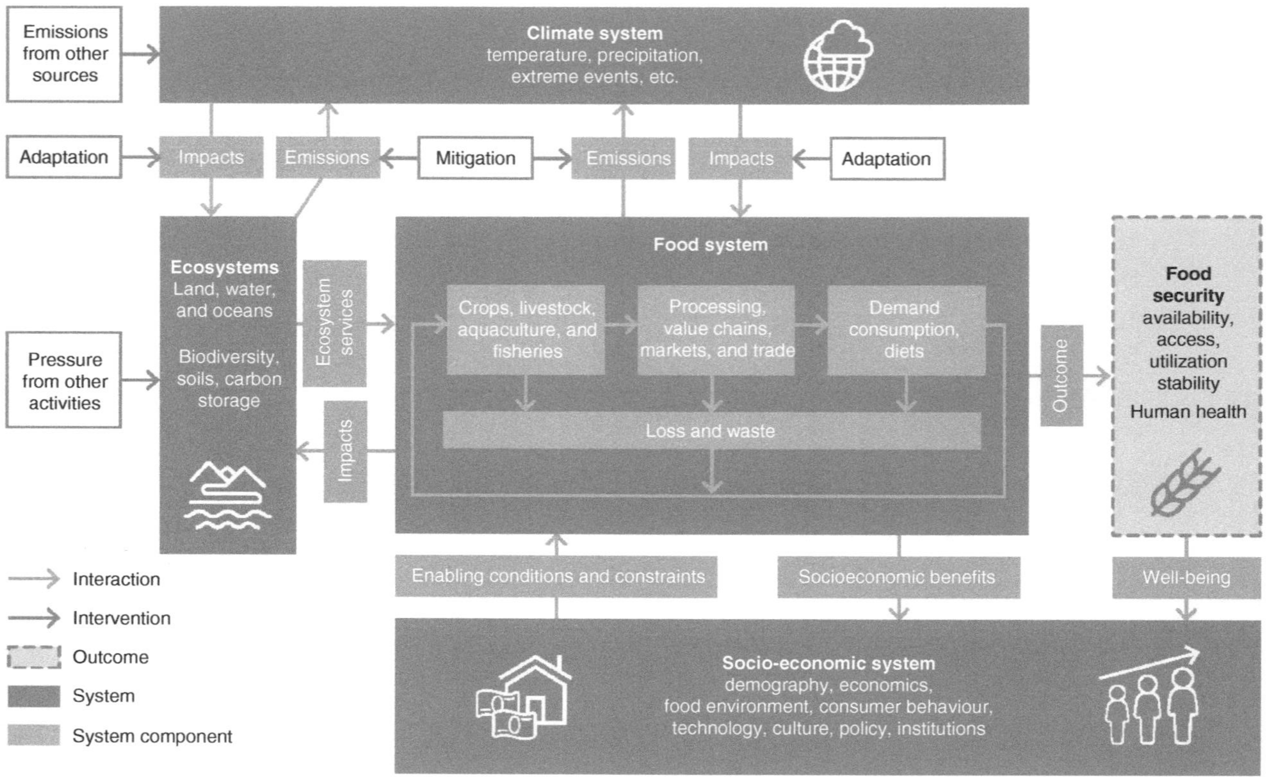

Figure 4.1 Interlinkages between climate and food systems, ecosystems (land, water, and oceans), and socioeconomic system. *Source:* Adapted from Shahzad and Amjad (2022).

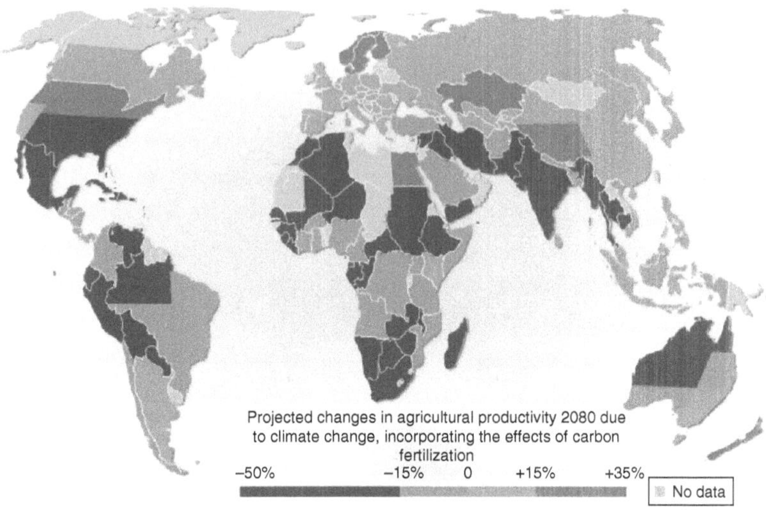

Figure 4.2 Glimpse of Africa and Asia-Pacific's climate change adaptation. *Source:* Adaptation to Climate Change (2017).

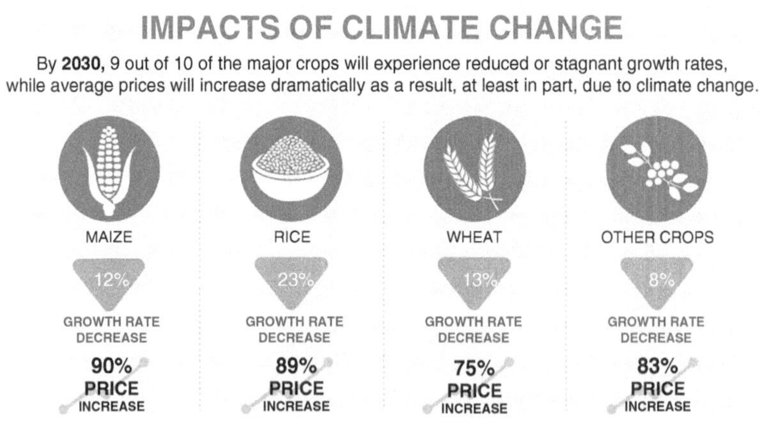

Figure 4.3 Impacts of climate change in Africa and Asia-Pacific. *Source:* Impacts of Climate Change (2017).

4.2.1 Climate Change and Food Security: Case Study of China

China became the second-largest economic power in the world after the United States as a result of the developing economic boom (Wang and Zhang 2022). The country has seen remarkable economic development over the past three decades, according to modern human history. In the

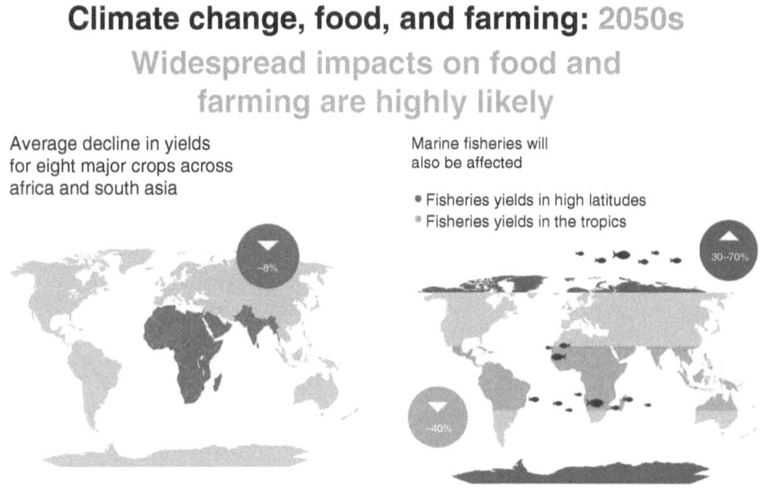

Figure 4.4 Climate change, food and farming: 2050 scenarios.

meantime, China's environment is becoming increasingly stressed because of its fastest economic expansion.

The four pillars of food security -availability, access, utilization, and stability -as well as their relationships, are predicted to be significantly impacted by climate change. And stoping hunger, providing sustainable food, and improving the quality of life are the core components of the sustainable development goals. There is widespread suffering due to the climatic elements' fluctuation as well as their changes, effects, and vulnerabilities. Any region in the world has already noticed the effects of changes in climate caused by global temperature and rainfall. However, climate change is already affecting our habitat, and food security and environment will make reaching the sustainable development goal even more difficult. Outcomes of climate change to the environment and societies are happening and widely distributed already; eventually, for sure, our nutrition is now standing in front of climate change, which is already transpired to be the most vital and challenging issue of our society. In the face of mitigation progress, now almost 800 million population are undernourished, under five-year-old children are estimated around 161 million to be nutrition-scarce, 500 million people are suffering from obesity, an unhealthy condition, and 2 billion are lacking proper nutrients they need to lead healthy lives. In such a rapidly developing world, increase in population, urbanization, and income are driving increased food production and consumption demand.

Figure 4.5a shows that more vulnerable region includes African countries. Fifteen out of 27 countries were from that region which makes up almost more than half of the proportion. Out of 109 countries, Mauritania and Mozambique showed the maximum national-level vulnerability. Out of many, one of the high dependencies of African countries was on fisheries. Fisheries were also a source to secure the food system, economy, and employment. Marine fisheries, for instance in Guinea Bissau, accounted for almost 59% of the total employment. Likewise, in Mauritania, 23% of its national gross domestic product (GDP) accounted for the economic value of marine fisheries. In the Democratic Republic of Congo, 38% (ave. protein per day intake) whereas merely 4.3 g of the above-mentioned protein's main source was fish.

High vulnerability was found in multiple Asian countries covering Thailand, the Maldives, Indonesia, the Philippines, Thailand, Vietnam, and

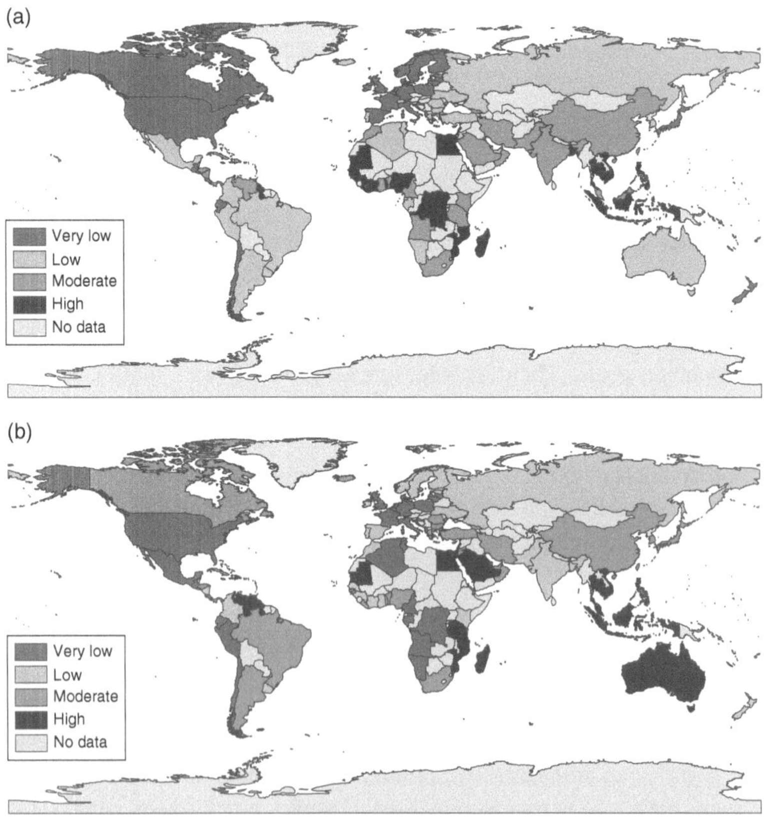

Figure 4.5 Food insecurity and climate change: Worldwide Scenario of Fishery Vulnerability. *Source:* Ding et al. (2017)/with permission of Elsevier.

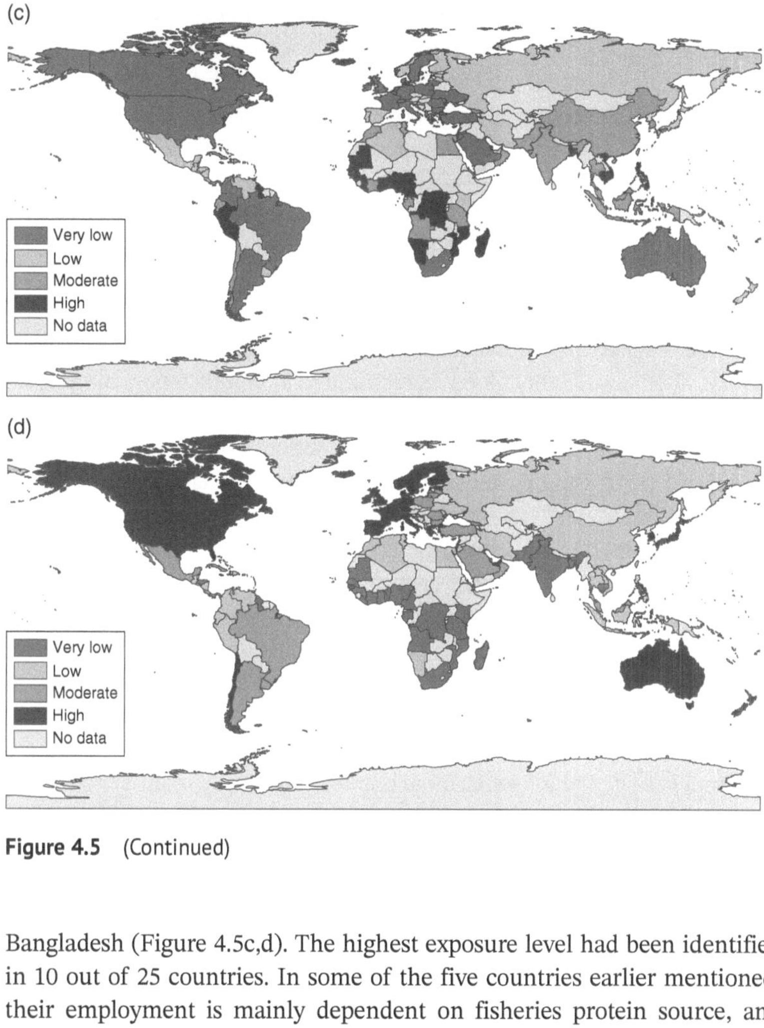

Figure 4.5 (Continued)

Bangladesh (Figure 4.5c,d). The highest exposure level had been identified in 10 out of 25 countries. In some of the five countries earlier mentioned, their employment is mainly dependent on fisheries protein source, and economy (Figure 4.5b). However, China is not among them and has low vulnerability due to its high adaptive capacity and low dependency level on fisheries. According to an estimation, food increment is a dire requirement until at least 60% by 2020 in order to meet the spiking demand of diet variety changes and growth in population, and it is practically obvious nowadays.

Food and agriculture have been concerned, though understated, importance in China's economic development nowadays. China has now become both the world's largest producer and consumer of agricultural products. Sustaining food security has always been one of the most important goals for government at all levels in China.

At present, agriculture in China is facing real challenges. Such as how to secure food supply and agricultural products to achieve the sustainability of agricultural development goal under the difficulty of limited resources and environmental suitability is the top challenge that must be overcome. Climate is one of the main factors in agricultural production among temperature and water shortage, which has seriously affected China's food security. Table 4.2 provides the overview of climate change impacts and vulnerability to the food security in China through socioeconomic dimensions. Several pieces of research have been conducted to study these relations of impacts drawn from the literature.

Climate change impacts usually influence four dimensions of food security: food availability, food accessibility, food utilization, and food systems stability. This serious issue will impact the human health, environment, food product, and food distribution which shall be seen in changing of purchasing power and market flows. Throughout the study of socioeconomic as well as biophysical adaptation approaches, tools to investigate the relationship are presented. The increased stress on food and agriculture production remained unconfined to food producer (the primary demander) in terms of availability of food. Nevertheless, increased mean of temperature relationship with crop failure shows the significance of this study. Potential for both aforementioned (socioeconomic and biophysical) is present in the available literature to suggest solution to climate change that affect food security. Meanwhile, it can be used to link agriculture production and climate, as methodology, in models and food security in different data.

Moreover, it is concluded that per capita food utilization of rural population is greatly affected by disaster in agricultural land. Accordingly, climate change will also impact the food supply, food demand, and ultimately utilization that may decrease huge amount of food fabrication. In addition, socioeconomic factor such as rural dweller's income would affect food utilization and in particular may cause the bad outcome due to inflexibility in food production and other risk in future (Orgeret et al. 2022). Consequently, increase in income of people will rise the purchasing power, like stimulation on goods demand, and this shows an optimistic link. Xiong said that it was clear that climate change had a huge impact on China food security. This research tries to measure such impact by using grain supply/capita, as it differs conferring to multivariate climate change circumstances (Figure 4.6).

Table 4.2 Food security and change in climate studies in China: an overview.

Studies	Author(s)	Methods	Issues researched
"Food Security, Food Prices and Climate Change in China: a Dynamic Panel Data Analysis"	Affoh et al. (2022)	The model and econometric approach	Estimate various data interrelation related to change in climate, prices fluctuations in food, and food insecurity by using econometric model.
"Increased crop failure due to climate change: assessing adaptation options using models and socio-economic data for wheat in China Environ."	Xie et al. (2022)	Crop and climate simulation use socioeconomic factors	Focusing on the crop failure and the importance of future variability in climate on different seasonal timescales that determine crop productivity using socioeconomic indicators.
"Climate change and critical thresholds in China's food security."	Li et al. (2023)	• The greenhouse gas emission scenarios • Crop estimationSynthesis • Socioeconomic scenarios	Contemplation of multiple indicators such as food trade internationally, growth in population, and technology development. Furthermore, quantification of crop production of China, a critical perspective.
"Future cereal production in China: The interaction of climate change, water availability and socio-economic scenarios."	Han et al. (2022)	Framework analysis based on socioeconomic scenarios.	Analysis and direct assessment of the climate variables and their varied effects based on socioeconomic scenarios.

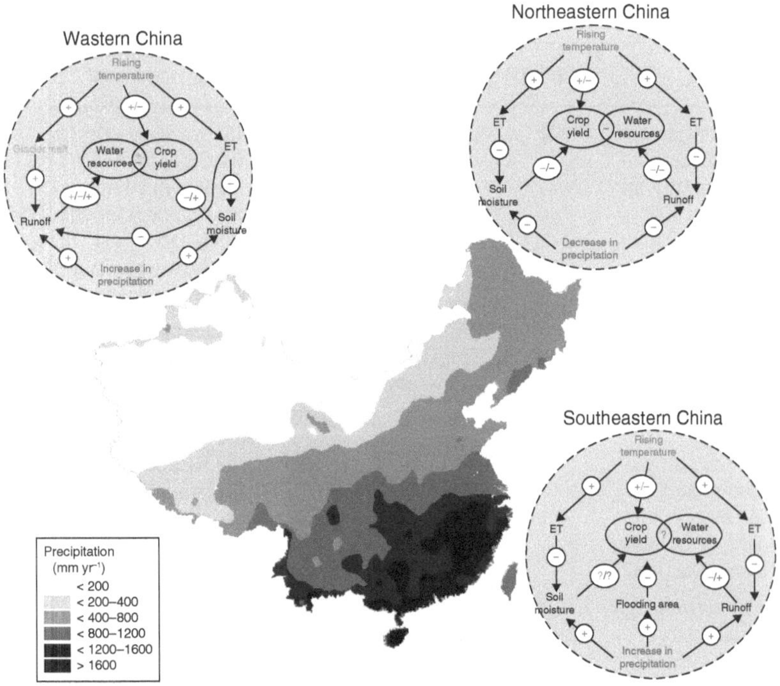

Figure 4.6 Climate conditions and variables in different parts of China.
Source: Piao et al. (2010)/Springer Nature.

4.2.2 Climate Change, Natural Environment, and the People of Myanmar

4.2.2.1 Background

Myanmar is sited between longitudes 92° 10E and 11E 101 and scopes 09° 32N and 28° 31N. This nation imparts a fringe to India, Thailand, Bangladesh, Laos, and China. The surface region of 676.552 square kilometer (sq-km) makes it the second biggest nation in Southeast Asia. Its coastline consists of 2832 kilometer (km) or over half of the whole eastern piece of the Andaman Sea and the Bay of Bengal. The nation is separated into eight physiographic districts. The zone incorporates the focal fields and Sittaung River, a precipitous region in the north, east, and west. Myanmar consists of three tropical seasons: from the month of November to February (winter season), from the month of March to April (summer season), and the rainstorm ranges from the month of May to October. Myanmar is, by far, the most part viewed as the nation and has wealthy characteristic assets: minerals, woodlands, prolific farming grounds with copious of Monson downpours, and rich marine assets. Myanmar is danger or cataclysmic event in the nation, being situated in the structurally dynamic Alpide Seismic Belt. Because of the fast development of populace, industry, and urban zones, different parts of the creating scene, natural corruption, and different dangers or man-made debacles, they are likewise on the ascent (MoNREC 2016).

Urban zones in Myanmar are similarly powerless against environmental change and outrageous occasions, as they are situated in the dry zone, beach front territories, and surge inclined or are encountering an expansion in the stream of water because of the increment of impenetrable surfaces. Individuals and resources in urban communities and towns have experienced past outrageous occasions and will experience the ill effects of these quick beginning fiascos, moderate beginning impacts of environmental change in future. For instance, Yangon has been hard hit by and has been assessed in littler urban communities, for example, Laputta needs to evaluate the misfortunes 33,000 for the township and extensive harm. In urban areas and towns which are confronting the outcomes of environmental change, a few gatherings are more defenseless against its effects, frequently poor and with the capacity to bring down fit. It is additionally likely that the impact of warmth islands because of high temperatures and densification of developed regions will bring about significant dangers to well-being and bearableness quality in borough zones. Aquatic deficiencies turn into a characterizing in both substantial and smaller towns as a challenge. Moreover, there is an increment in costs of administrations in bigger urban areas, for example, Yangon, Mandalay, and Nay Pyi Taw, with ensuing medical problems.

Populace development alongside monetary development will expand the interest in vitality and different administrations in urban regions, which thus prompts an expansion in ozone-harming substance outflows, for example, structures, transport, and industry. Structures with insufficient vitality proficiency norms will advance crisp structures a great deal of stock of ozone harming substance outflows in Myanmar later. Political change alongside monetary development demonstrates a huge increment in the vehicle division, including the car development—fewer than 1 million vehicles enlisted almost 4 million of every 2012 in 2004—and keep on growing as the financial chances of family units to increment and differentiate the gainful area.

4.2.2.2 Climate Change and Myanmar, A Societal Perspective

The unfavorable effect of environmental change cuts crosswise over segments and social orders. In this way, the effect is felt in one part and different segments of society all in all. For instance, the negative impacts of environmental change on agribusiness will lessen its commitment to GDP development and the occupation of little local and farming work strong. In Myanmar, 2010 elections heralded a partial shift to a quasi-civilian rule and economics, and the 2016 national elections signaled the expansion and deepening of economic liberalization, notwithstanding the military's continued entrenchment in the state and society (Aung 2022). In general, extractives development has been the preferred catalyst for economic expansion. Foreign investors and domestic businesses founded primarily by military insiders have long attempted to control cheap labor and natural resources, but they have encountered at least two institutional and political barriers: the absence of enabling legislation and the bloody conflict with the ethnic armed groups whose territories most of the resources are found. There were three main political dimensions of conflict in agricultural cultures prior to the age of climate change politics and the present global land rush: working people versus the state, working people versus businesses, and working people versus working people. People involved in conflict do not join conflict resolution procedures and processes as random individuals separated from their socioeconomic and political social class ties. Instead, these conflicts and the attempts to resolve them are enmeshed in wider social relations.

These drivers may prevent Myanmar's capacity to accomplish its improvement goals, as characterized in their plans and secured by the Constitution. As watched, given that occupations and the economy depend on atmosphere touchy parts and are situated in atmosphere delicate zones—environmental change will represent a tremendous test to accomplishing comprehensive financial and social improvement (IPCC 2014). For instance, around 85% of the country populace depend for their subsidizing on atmosphere touchy divisions of employment, and many individuals are packed in regions influenced by environmental change, incorporating into the delta district and seaside zones. The rising ocean level stance major issues for beach front networks and vocations, including seaside biological systems and environment administrations. Rice output that also can be pronounced as yield is expected to lower down due to changes in environmental situations in Myanmar. Managing nourishment generation and keeping up sustenance in this manner are outcomes that Myanmar would confront including Asia-Pacific (Thein 2015).

While financial development, social improvement, and ecological supportability will be influenced by antagonistic effects of environmental change, troublesome advancement direction, especially vitality and asset extreme improvement, could bring about expanded debasement of characteristic assets and increment of GHG discharge. GHG commitment from the business and development divisions joined was 10% out of 2012.

Yearly coal generation is required to build altogether to 5.6 million tonnes (mt) by the year 2031 and will prompt an increment in GHG emanation. Environmental variations can likewise influence social improvement, prosperity, well-being of humans, and training. Effects may be borne by officially helpless networks and underestimated districts in the first place and will in this way undermine endeavors went for adjusted and comprehensive improvement. There is in this way a need to react to environmental change. Figure 4.7 gives few effects on indigenous habitat and the public of Myanmar and diverse helpless areas of Myanmar.

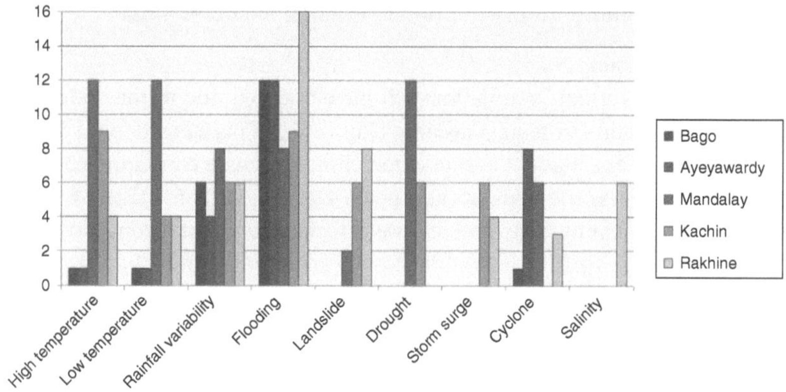

Figure 4.7 Climate change impacts on the natural environment and the society. *Source:* Adapted from MCCA (2018).

The Myanmar government has the chance to create in a feasible way that is solid for the economy and atmosphere. Notwithstanding the advancement of changes in the vitality area system to lessen ozone harming substance outflows, Myanmar incorporates changes in farming, ranger service, and waste divisions. Governments are regularly hesitant to actualize approaches to diminish the effect of atmosphere, since they trust that the expenses are too high. The securing of the innovation required for viable strategies for development and creation can be troublesome and exorbitant.

Creating nations, with the assistance of created nations, have the chance to create in an economical way. World Bank, ADB, UN, and ECDO (Economic Cooperation and Development Organization) alongside other nonadministrative associations, universal associations, and privately owned businesses give a strategy to creating nations to procure the innovation required for innovation exchange and financing manageable improvement. This permits nations, for example, Myanmar, to develop and meet the goals suggested by UN through SDGs in the growth process.

4.2.2.3 Concluding Remarks

Following are the few recommendations based on the earlier-explained case study of Myanmar. Policymakers should add an improvement in demands in a maintainable manner in a sustainable process for the betterment of earth and economy. They also need to advance training and open mindfulness for environmental change and build limit with respect to flexibility and adjustment on environmental change to partners. The government needs to create environmental change strategy and techniques for environmental change administration in Myanmar. Decision-makers should plan to have networks and monetary areas ready to react to and recuperate from atmosphere-initiated calamities and dangers and assemble a sound society.

4.2.3 Social Vulnerability to Climate Change in Urban India

4.2.3.1 Background

By 2050, India's urban population will have doubled due to the country's rapid urbanization in the next decades (Jain 2021). This transition is special because it will take place at a time when climate change consequences and pressure to address them are at their peak. Thus, the urbanization of India will have significant effects on local welfare, environmental conditions, as well as climate mitigation and adaptation (Pandit 2021). India has a unique chance to secure low-carbon, resilient, and egalitarian urban forms for the long future because the majority of its metropolitan areas are still under construction. Powerlessness turned into change, especially in the environment, is characterized as notch to which a city or a social framework is influenced by the negative impacts of environmental change. Defenselessness additionally considered an element of versatile limit, affectability plus presentation. Environmental variation considered as the vital problem and been an erratic impact on framework of the society or on social aspects of the society. Subsequently weakness of the society depicting into environmental fluctuations has gained the attention of numerous specialists to contemplate its effects. It identifies the presentation of individuals to the impacts of environmental change and their weakness to it including the close-by zones that are straightforwardly influenced by the delicate territories (Komali Yenneti et al. 2016). In this manner, social helplessness gives a more exact framework for think about instead of physical powerlessness (Komali Yenneti et al. 2016).

India is also known as Hindustan, a nation of differed topography plus shifted disclosures to environmental dangers. Geographical location and surrounded territories made this nation a significant piece defenseless against the different perils of environmental change. India is one of the 170 nations that are to a great degree powerless against environmental change impacts, distinguished by some quickly developing financial matters throughout the following 30 years. In India, a loss of life of 3,00,000 individuals was seen in the violent wind that happened on 25 November, the super typhoon of Orissa on 29 October and the 2001 Bhuj tremor are still recognized as real occasions in the most affected history of Indian's occupation. Water and farming are the most powerless segments to environmental change according to the ongoing reports of IPCC.

Coastline of India is spread over 7500 kilometers (kms) altogether and 5700 km on terrain. This line is extended from most western side of Kutch bay to the world famous forests and mangroves of Bengal or West Bengal known as (sunder bans) (Krishnan and Lakra 2013). The dissolving of Himalayan icy masses directly affects the accessibility of water assets in India. Precipitation examples will likewise be influenced, bringing about interruptions to edit creation and horticultural patters that specifically affect the low-pay gatherings. Analysts proposed differing hypothetical models and charters to evaluate social powerlessness of multiple land settings. Under these structures, presentation, versatile limit, and affectability are three noteworthy social pointers. Presentation alludes to the levels of social framework introduced assortments of important climates. Versatile limit is as far as possible to respond to an atmosphere aggravation and adjust to the impacts of the agitating climatic rates to survive. The nonappearance of versatility, with high presentation and affectability, can assemble the defenselessness of a structure. Affectability is how much a framework is influenced in the structure (Komali Yenneti et al. 2016). One of the helplessness considers directed by Indian scientists utilizing comparative markers of introduction, versatile limit, and affectability on the eastern bank of India observed the net impact to be negative in 10 beach front areas.

4.2.3.2 Natural Calamities and Adaptability

The impacts of characteristic catastrophes and environmental change differ generally by state, district, and assembling. Even though social factors, for example, wage and age, do not pick a person's identity when hit by cataclysmic events, they do pick a populace's capacity to plan, react, and adjust when disaster strikes. Undeniably, it has been found that environmental change and social vulnerability have been directed at different regions with moving social conditions in fluctuating social circumstances and the change of the area. In this regard, the Social Vulnerability Index (SVI), basic gadget, has been proposed to estimate the vulnerability of the overall population to environmental switch to set up the adaptability processes. Such research relies upon a relevant investigation coordinated in India to the extent of environmental change and social vulnerability concerning the urban networks in India.

A marker-based approach estimating social defenselessness has ended up being effective in the present year. While broad research on helplessness to environmental change in India has ascended consistently, its overwhelming piece has focused on assessing powerlessness as a whole. The current survey is surprising and quite dissimilar to the past audits, and this provides a perception of social helplessness over a time allotment and city populace. The social powerlessness in India will spread over a spatial scale with wide monetary assortments and differences in physical and social conditions. In view of the data analyzed by the scientists (Komali Yenneti et al. 2016), mapping was finished utilizing GIS apparatuses demonstrating the varieties in the three pointers and looking at the 29 conditions of the nation. Figures 4.8–4.11 demonstrate the varieties of the three pointers. The different file esteems are signified in view of whole urban area of India. In the middle, total qualities above it are viewed as high and underneath it are low (Komali Yenneti et al. 2016).

4.2.3.3 Exposure

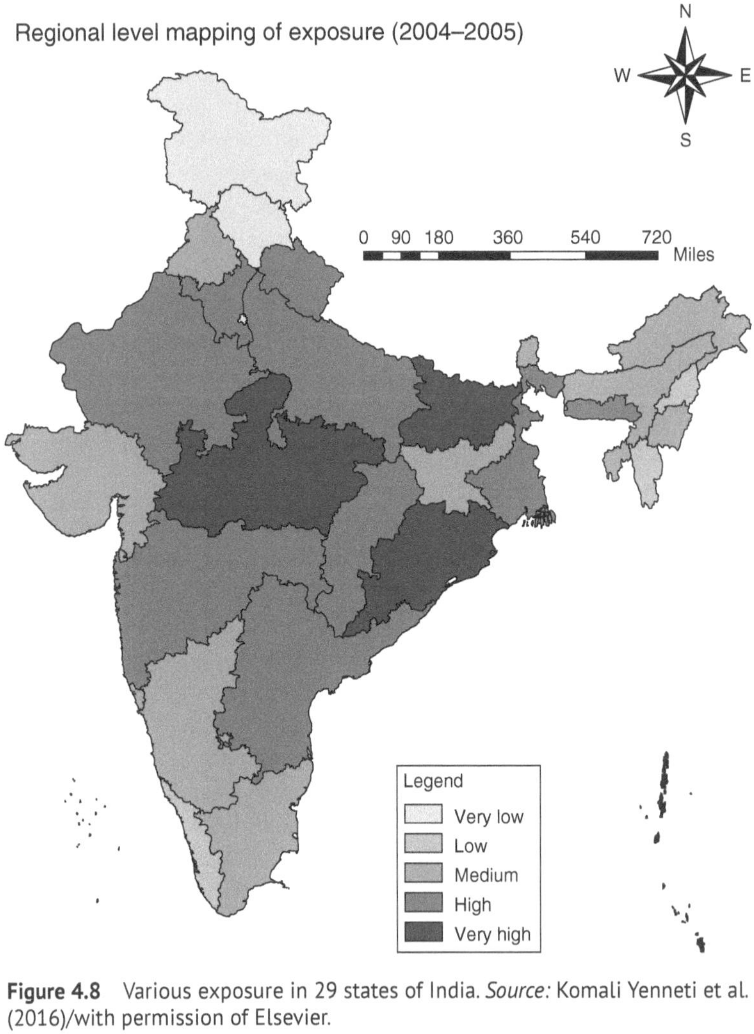

Figure 4.8 Various exposure in 29 states of India. *Source:* Komali Yenneti et al. (2016)/with permission of Elsevier.

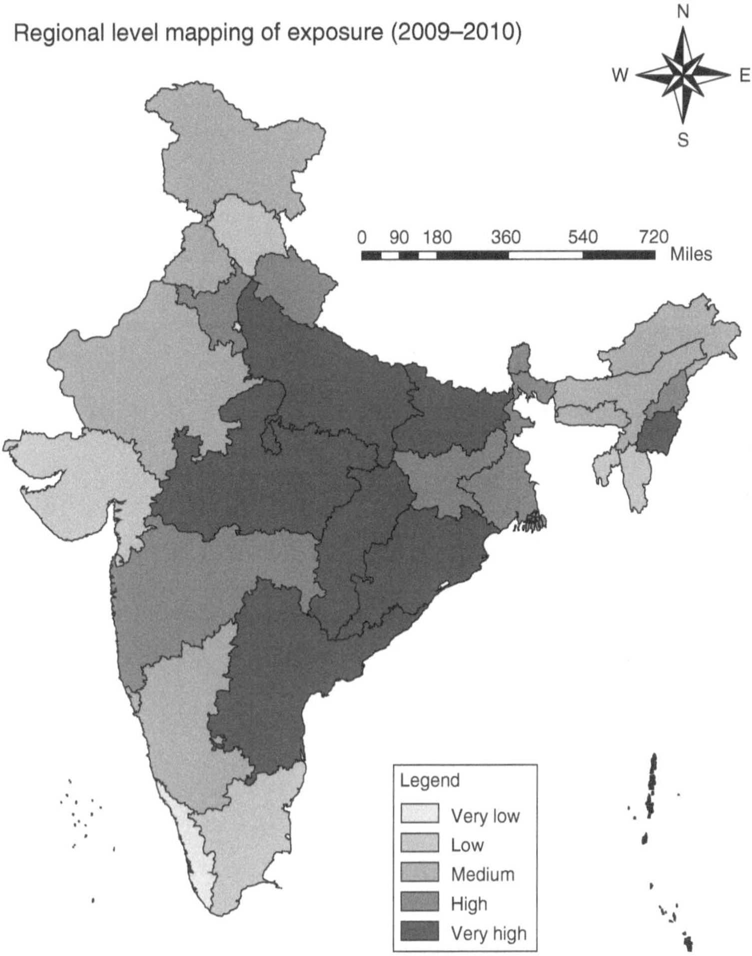

Figure 4.8 (Continued)

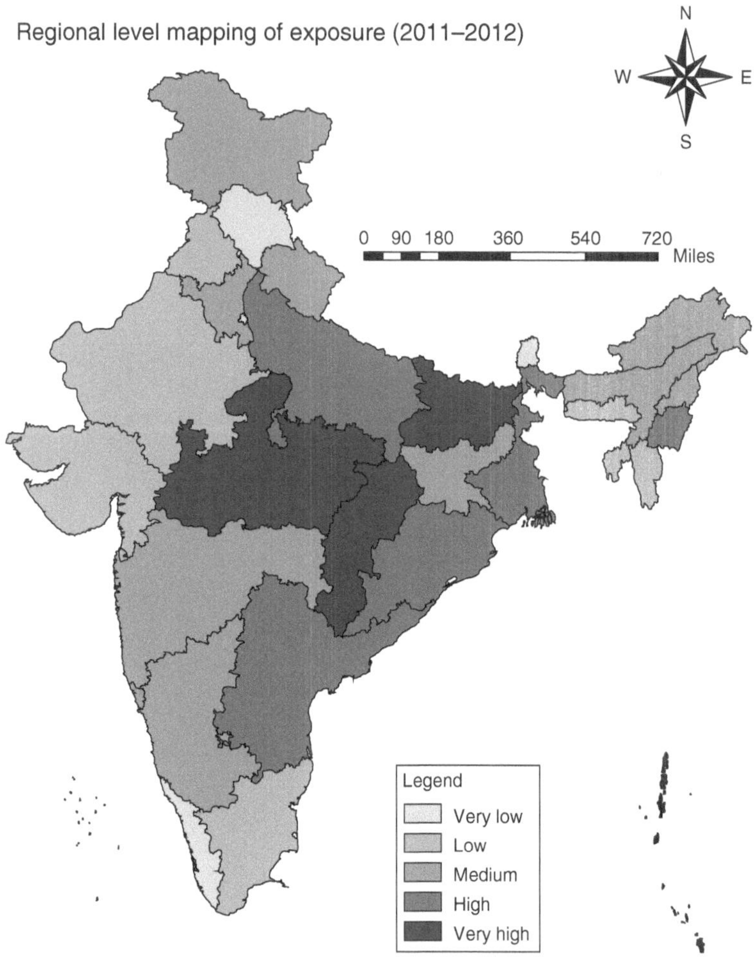

Figure 4.8 (Continued)

4.2.3.4 Sensitivity

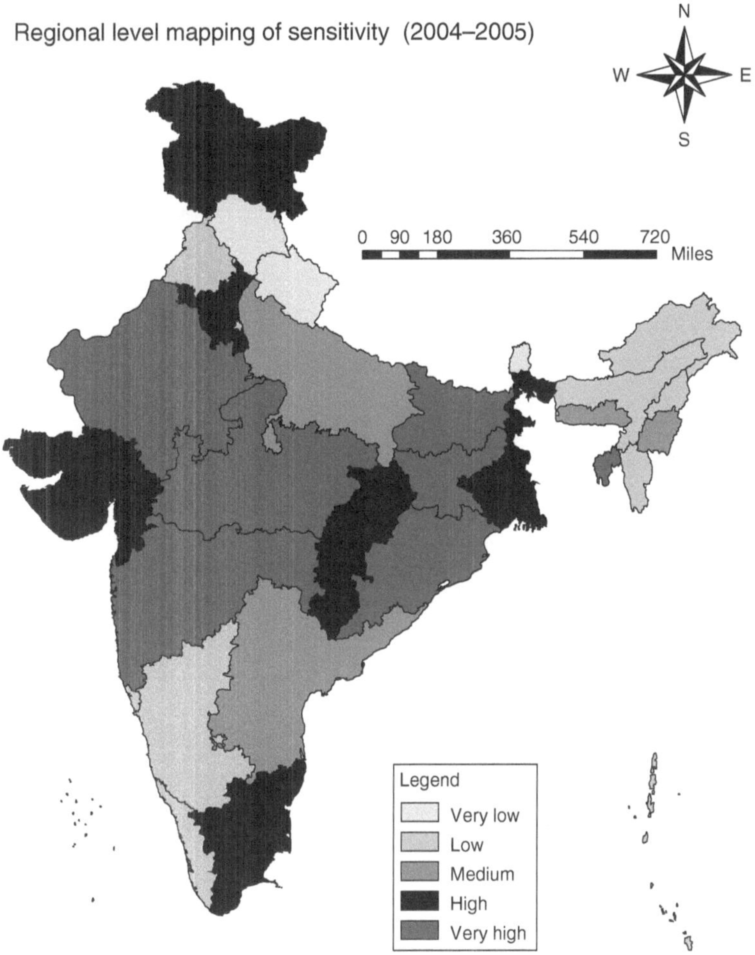

Figure 4.9 Variations in the sensitivity in 29 states of India. *Source:* Yenneti et al. (2016)/with permission of Elsevier.

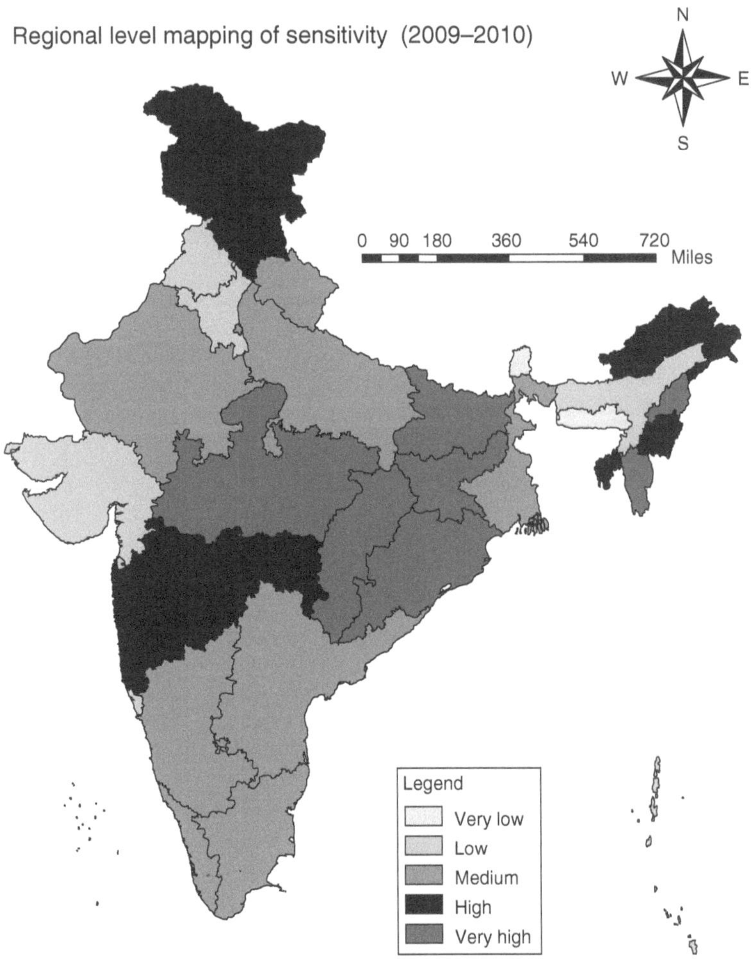

Figure 4.9 (Continued)

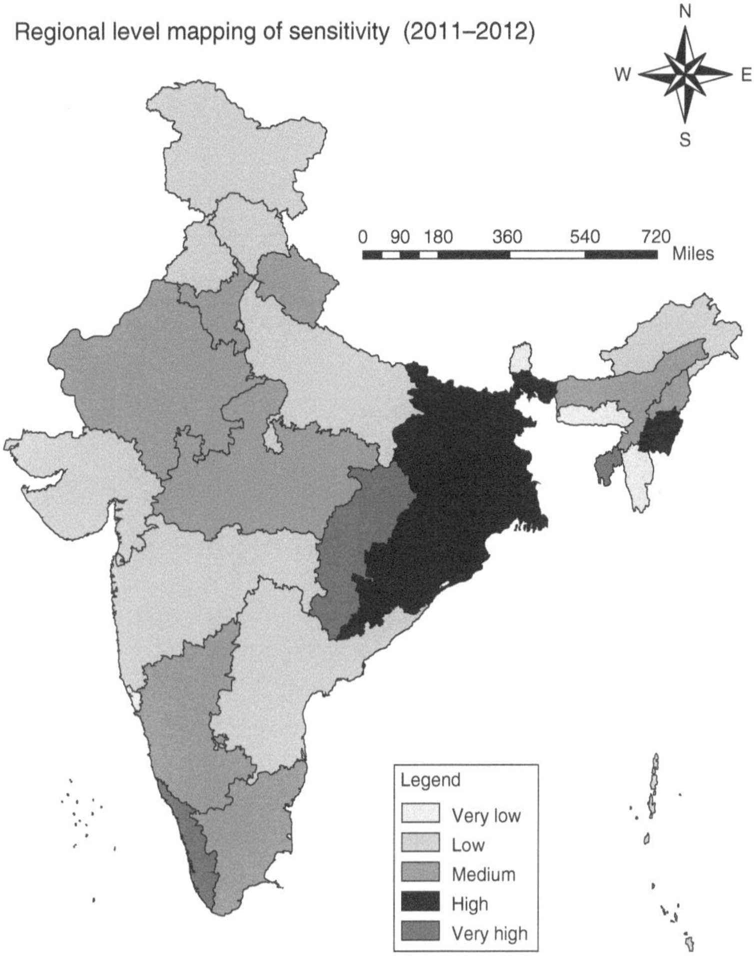

Figure 4.9 (Continued)

4.2.3.5 Adaptability

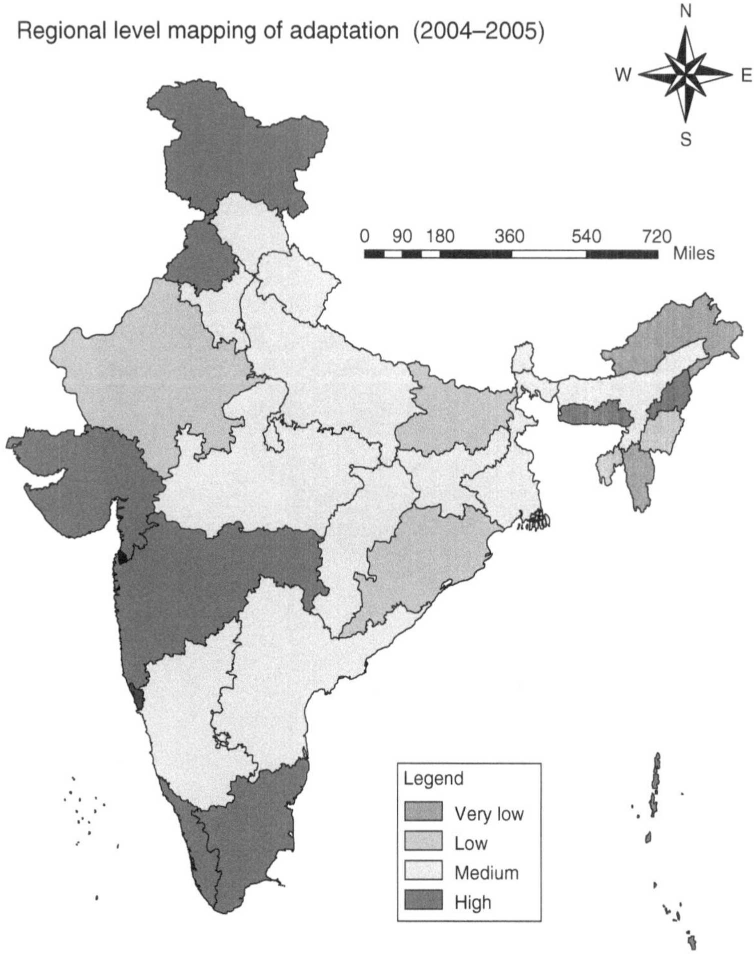

Figure 4.10 Variations in the adaptability in 29 states of India. *Source:* Komali Yenneti et al. (2016)/with permission of Elsevier.

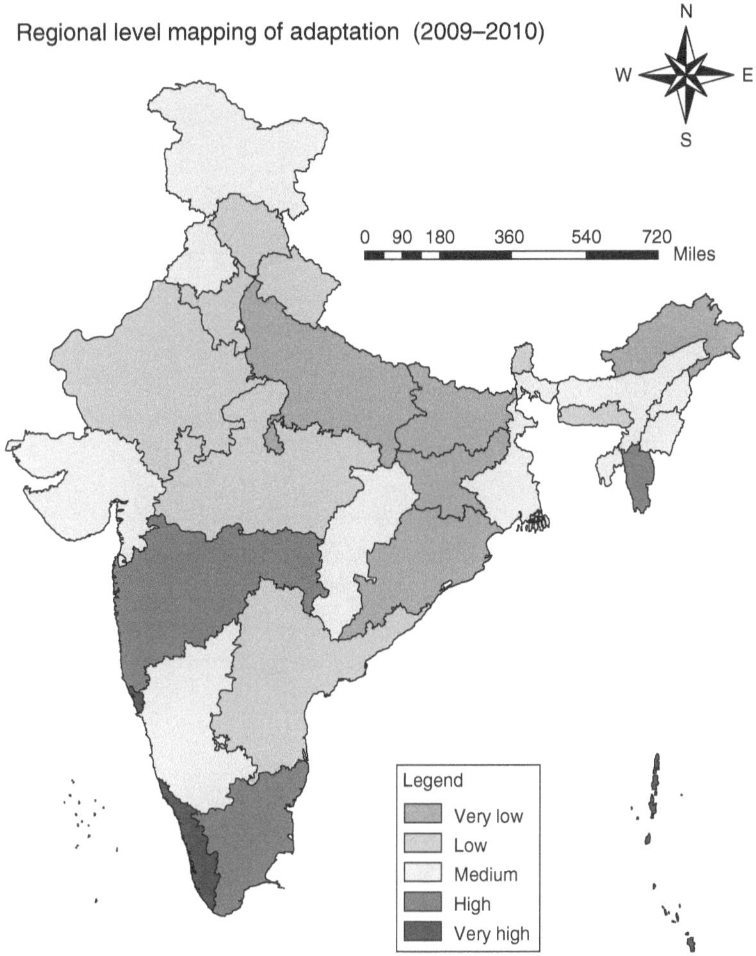

Figure 4.10 (Continued)

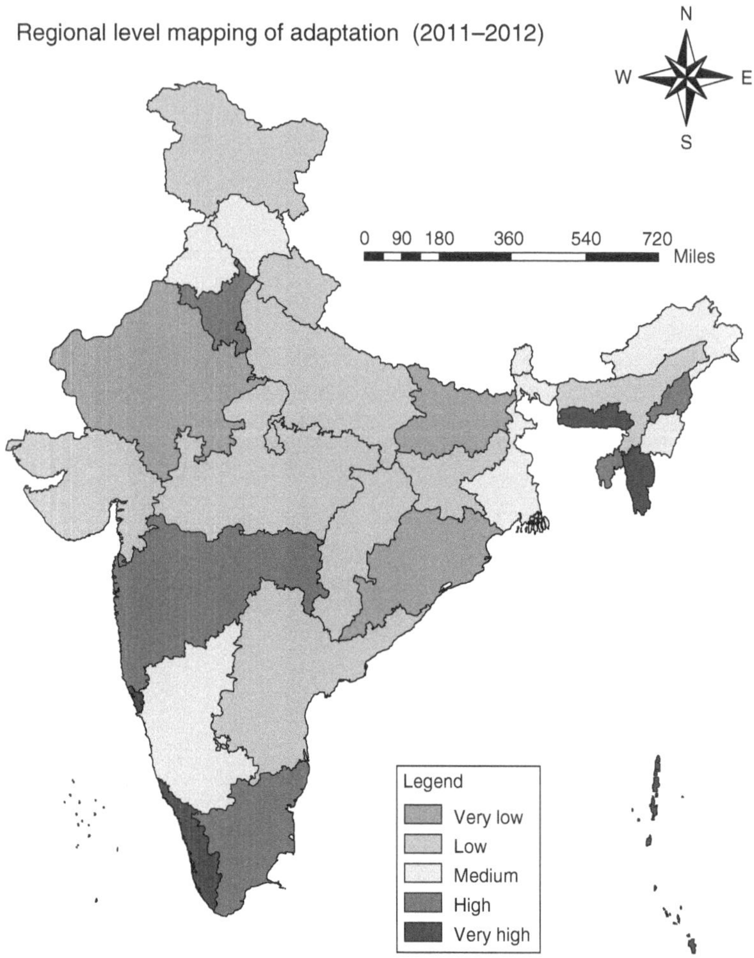

Figure 4.10 (Continued)

4.2.3.6 Vulnerability

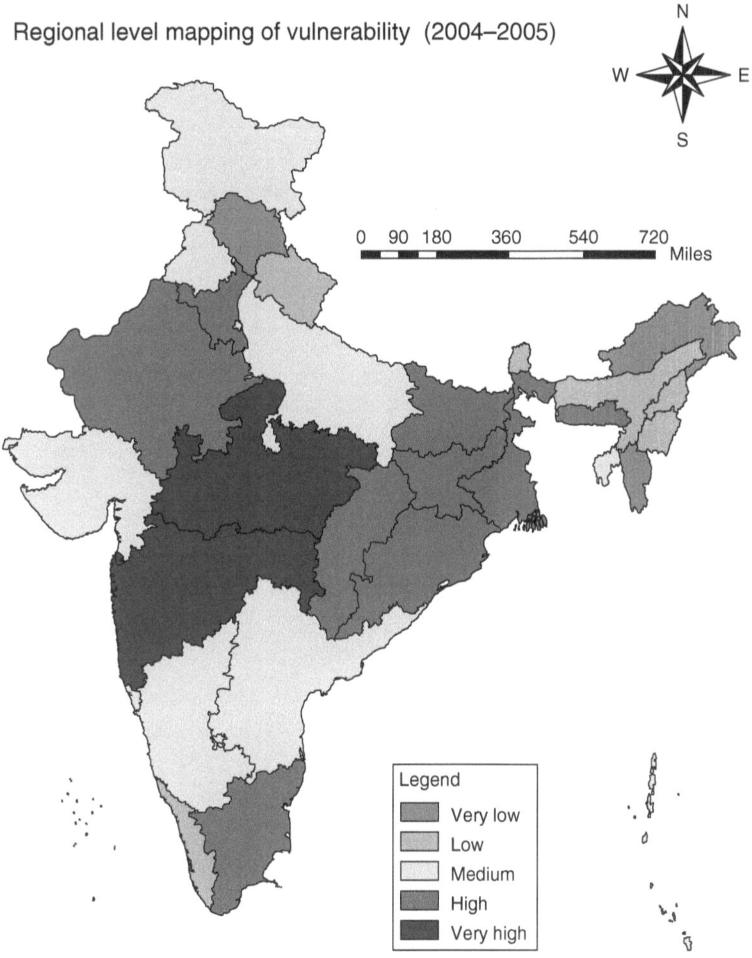

Figure 4.11 Vulnerability variations in 29 states of India using CUVI.
Source: Komali Yenneti et al. (2016)/with permission of Elsevier.

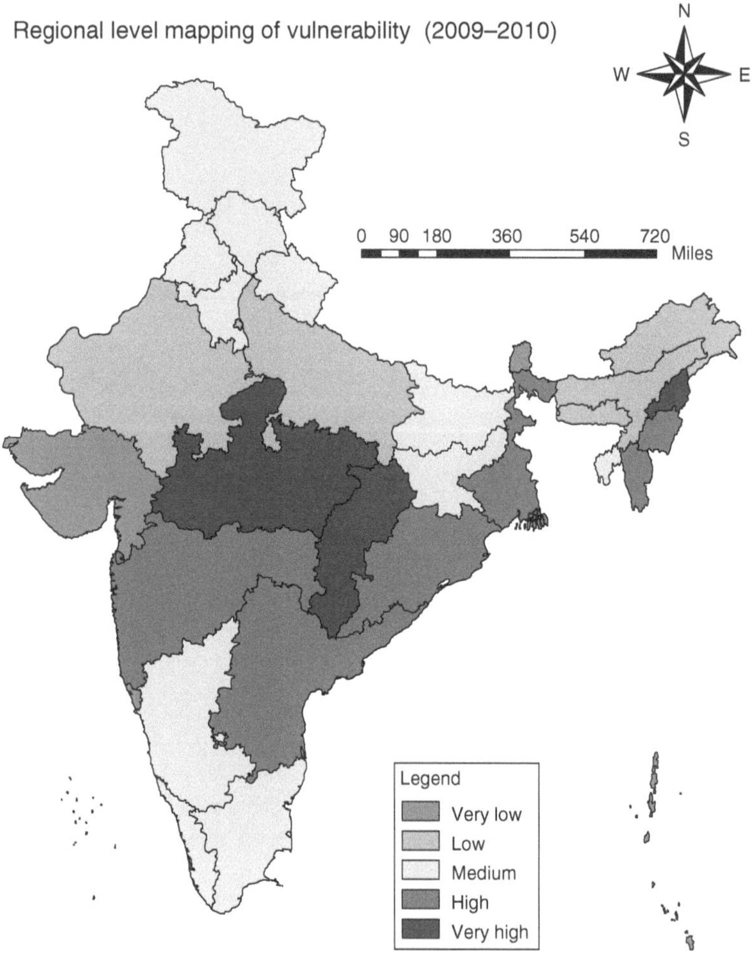

Figure 4.11 (Continued)

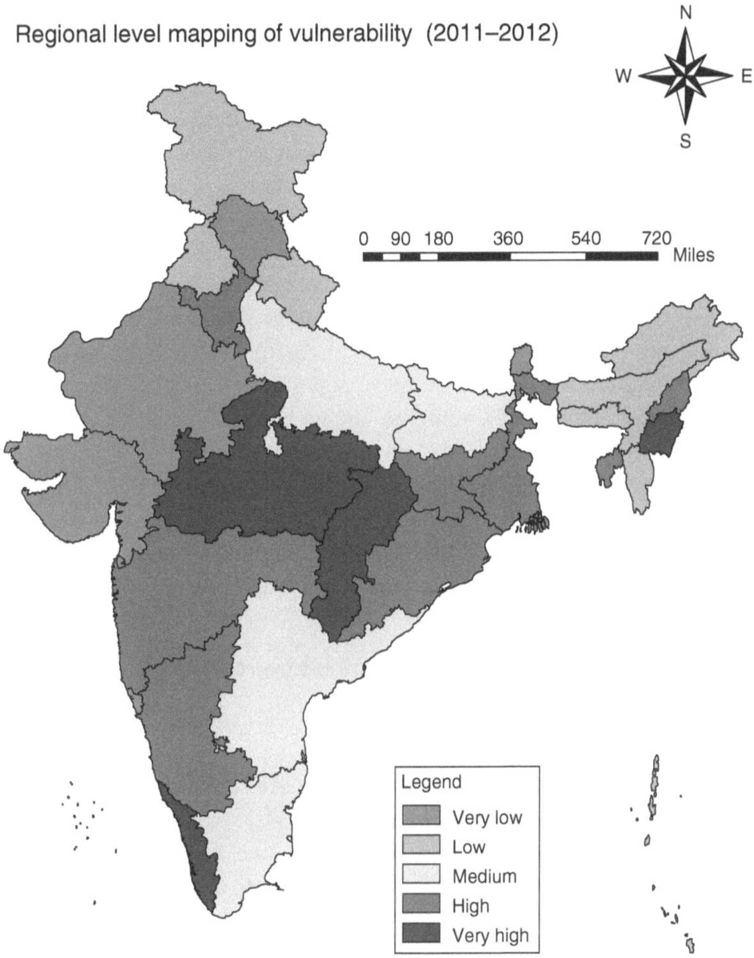

Figure 4.11 (Continued)

4.2.3.7 Concluding Remarks

The aforementioned case study provides insights on the vulnerable population in India that benefits the policymakers and the government planners, if used in effective manner. To begin with, it can take in the most powerless spots in the nation. Second, it can encourage metropolitan government and the people to plan for satisfactory catastrophe administration groups and strategies. Without the knowledge of the current situation of vulnerability, it is very difficult to address the issue. The knowledge of strengths and weakness of different areas provides better efficient management in order to cope up with the issues of change in climate. The CUVI[2] method helps to get a brief idea regarding the same (Komali Yenneti et al. 2016). The present system of calculating vulnerability can be used to significantly enhance the current vulnerability situation of the country to climate change. A standard calculator of vulnerability may be included in all thought-provoking processes of policymakers so that the policymakers have a better grasp on how to handle the situation in times of disasters. Some advancements could be made to cope with the issues of change in climate so called climate change through mixed approach, i.e. qualitative, and quantitative measures soon.

4.2.4 Climate Change Impacts on Agriculture-related Food Security in Pakistan

4.2.4.1 Background

Losses in crop yields in the agricultural crops are somehow aftereffects of climate change which shows through the empirical studies of last few years utilizing multiple models and techniques (Bandara et al. 2014). IPCC (2013) claimed that an increase in temperature of 2°C above preindustrial levels leads to fatal risk of sever climate change impacts and global temperature rate is increased from 0.2 to 0.3°C. The present level of CO_2 concentration is 430 ppm, and it is rising almost annually 2 ppm as per estimations.

Economy of any country depends on its major growth sectors. Developing countries have agriculture as their major economic sector, and fluctuations in their growth have heavy impact on the economy of a country. Pakistan is agro-based country, and industry is also an important sector now where every kind of raw material is provided by agriculture sector, so change in agriculture sector affects the whole economy. Climate is a major factor as humidity and precipitation rate affect the production and consumption patterns of agriculture; in other words, it is a major variable which has great

2 Composite Urban Vulnerability Index.

influence on agriculture. During last two and half decades, change in the climate of Pakistan has drastic effects on all the sectors of agriculture in different ways. There is limited work done in this regard; moreover, its impact on subsectors like livestock, fisheries, and poultry are very rarely known so far.

World projection about populace demonstrates that constantly 2080 total populace end up 9.2 billion; nowadays, it is 5.9 billion, so nourishment request is anticipated to increment by around 300% continuously 2080 because of higher populace, and this ascent is probably going to make an opposite corresponding connection between sustenance free market activities. The confirmation of various examinations demonstrated that agribusiness area contributes for a substantial offer in their gross household items (GDPs) in creating nations (Siddiqui et al. 2012). Hence, the advancement of the economy cannot be accomplished without enhancing the agribusiness part. Pakistan is situated in South Asia on 75.5 East longitudes on 240° to 37° North. Precipitation and mugginess are two key factors in the yield of agribusiness wares.

Economic Survey of Pakistan shows that contribution of GDP in agriculture sectors is almost 21% in recent reports (Pakistan Economic Survey 2014). Pakistan is basically an agro-based country, and climatic situations are very drastic in Pakistan as climate is humid. The agriculture sector of Pakistan consists of various subsectors, including farming, fisheries, poultry, dairy, and many others. Major staple crops of Pakistan's agriculture are cotton, rice, wheat, maize, and sugarcane. *Kharif*[3] and *Rabi*[4] are two agricultural seasons in Pakistan. Agricultural productivity is heavily influenced by the climate and weather fluctuations. The agricultural segment stands as the significant backbone of the economy of Pakistan which is in charge for 67% almost of the incomes associated with this sector. Pakistan is a large and populated country of which almost 32% are poor being below the poverty line. In this case, major stay of livelihood and more than 47% of people directly or in directly engaged in agriculture sector in Pakistan, so climate change has drastic effect on agriculture and directly affects the income of rural people (Pakistan Economic Survey 2015). Pakistan is undoubtedly highly vulnerable to change in climate as environment of Pakistan is most suitable for disease attack and agriculture markets are more volatile in price fluctuations. Hence, this case study aims

3 Grown in summer season (usually from month of May to month of November) is called Kharif season, and its major crops are rice, corn, and cotton.
4 Grown in winter season starts (from month of November or month of December ends up to month of April), significant yields include as wheat, barley, and millet.

to analyze the impacts of climate change on Pakistan's agriculture in the context of flood damages mainly.

As indicated by the conference of parties (CoP 21st) gathering of Paris, Pakistan is incorporated into top three very presented nations to environmental change out of 10 (Rehamn 2015). Rising temperatures, extreme downpours, dry seasons, and creation misfortunes in agrarian segment are relied on to expand more in Pakistan because of environmental change. Pakistan has been bearing the effect of environmental change without having earlier learning of it as no genuine consideration is paying to environmental change and agriculturists are uninformed of it. It is specified that environmental change impacts are more in bumpy regions as downpours would escalate in the northern area of Pakistan. The overwhelming and sporadic downpours caused visit surges in Pakistan.

4.2.4.2 Climate Change, Floods, and Food Insecurity

The causal relationship between climatic fluctuations and practically every element of life has been extensively seen and verified by humanity. Severe losses to human life and the environment have been brought on by climate catastrophes, including floods, storms, and earthquakes. Extreme climate dynamics in the last few decades have led to a rise in the occurrence of riverbank erosion and subsequent flood catastrophes. Floods are a topic of research in the hydrology field and are very important to the fields of agriculture, civil engineering, and public health. Due to the current climate circumstances, the world has been seeing a wide range of particular and serious obstacles in meeting human beings' fundamental requirements. Particularly, the activities and changes affecting the climate are strongly related to our food system. In the long run, these climate variables are impacting our nutritional demands and escalating food insecurity due to our own activities (Fanzo et al. 2022).

During last two decades (2010–2022), Pakistan has faced two mega flood disasters. In 2010, flood severely hit northern province of Khyber Pakhtunkhwa (KPK) of Pakistan and then Punjab and Baluchistan. In 2022, almost 1/3 of the total area was under flood. Pakistan has regular flooding just like the rest of South Asia. Pakistan has a long history of devastating floods (Atta-ur-Rahman 2010). However, the disastrous flood of 2010 is recognized as the worst flood of the century. Along with severe flooding (2010) in the Indus that inundated a significant portion of the D. I. Khan district, the province's rivers especially Kabul, Swat, and Panjkora suffered very high flood flow along their courses. In several places, the floodwater was more than 6 m. The rural and urban communities were completely submerged by this significant flood depth. As a result, the discharge at several monitoring sites exceeded the high flood level by a wide margin. According to a field study, none of the respondents had ever encountered a flood of this size and scope in their lives due to the volume and scale of the 2010 flood. Depending on where the flood region was, the time it took for the water to drain ranged from three days to two weeks. According to the most recent figures, 1961 people have perished and around 1million homes have been impacted (Ali et al. 2018). This affected almost 20 million people and devastated infrastructure, crops, and houses, leaving millions of people at risk for starvation and waterborne illness. Between 1200 and 2200, people are estimated as dead in total, and 1.6 million dwellings were either destroyed or damaged, leaving an estimated 14 million people without a place to live.

In 2022, according to the country's National Disaster Management Authority, initially 937 people have perished due to heavy rain and flooding since mid-June in the South Asian nation (NDMA). On July 22, Pakistan's hilly northwest area had recorded monsoon rainfall that led to flash floods in the provinces of Khyber Pakhtunkhwa, Punjab, and Baluchistan. Unprecedented rainfall overpowered flood barriers, destroying bridges and highways, and inundating vast tracts of land. The floods were referred to as "unprecedented" and "the biggest humanitarian tragedy of this decade" by Sherry Rehman, the Federal Minister of Climate Change. Rehman noted that Pakistan was experiencing its eighth monsoon cycle, as opposed to the three to four cycles of rain the nation typically experiences. "Super flood torrent statistics are startling." According to Rehman, the neighboring province of Baluchistan, which is mostly cut off from electricity, gas, and the internet, has asked for 100,000 tents, while the southern state of Sindh, which has been severely affected by the flooding, has asked for 1 million The BMJ quoted public health expert Malik Muhammad Umair as saying, "With most of the flood affected areas having limited access to safe drinking water, the biggest threat is from waterborne diseases such as diarrhea, typhoid, and cholera." Malik Muhammad Umair was a member of response teams led by an international nonprofit organization after the 2005 earthquake and floods in 2010. Furthermore, as Pakistan has already been dealing with a high number of instances of malaria and dengue, the increased mosquito breeding brought on by stagnant water might have a similar effect. Totally, 116 of Pakistan's 154 districts have been impacted by the floods, which were brought on by the intense monsoon rains that began in mid-July. So far, more than 1162 persons have been passed away (WHO 2022).

Table 4.3 shows some statistical data on losses and damages due to devastating flood in 2022 in Pakistan. It is evident to see that more than 6.6 million population has been affected directly and indirectly, while

Table 4.3 Damages/losses due to flood 2022 in Pakistan.

Area affected	Crops	Population affected	Houses destroyed	Livestock deaths	Reported deaths	Schools affected
3.6 million acres	304,475 acres (destroyed)	33 million	218,000	794,000	1,033 (deaths)	17,566
–	2 million (impacted)	–	452,000 (damaged)	–	1,343 (injured)	–

Source: FAO (2022).

3 million is affected directly (FAO 2022). Not even a single sector remained safe during this natural calamity. People, livestock, crops, schools, roads, bridges, telecommunication, and transport systems have been destroyed, and it is expected that it will take many years to rehabilitate the damages. Figure 4.12 presents comparative view of flood-affected areas of Pakistan in

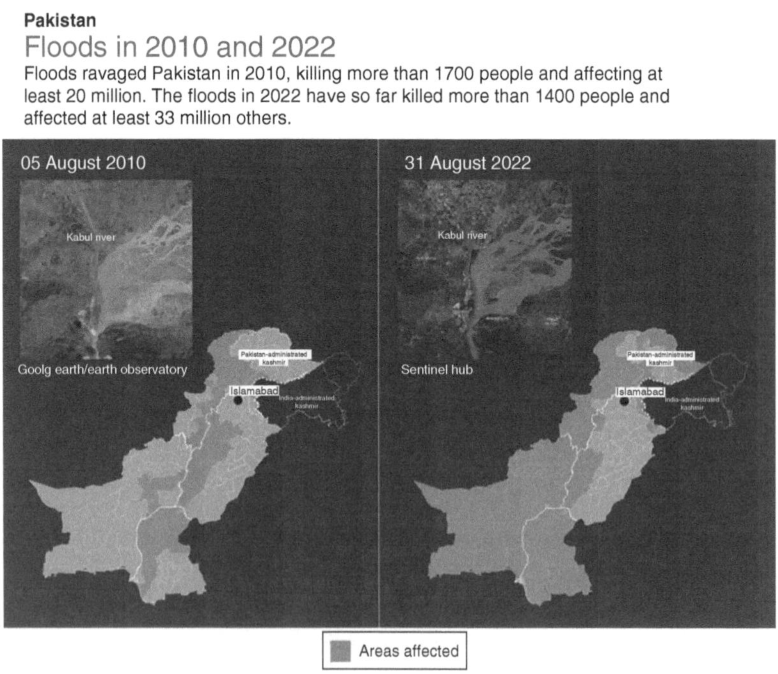

Figure 4.12 Comparative view of flood-affected areas (2010 and 2022). *Source:* Al Jazeera Media 2022.

2010 and 2022. That means, almost half of the population will face food availability issues in next few months due to crop losses at significant scale. This flood-led crop loss will not only a damage to the flooded areas but also to the whole nation. That alarms for a serious food availability and accessibility issues not only in affected areas but also in the whole country. As Pakistan is already under huge pressure from foreign debts due to political instability and low productivity, this will appear as another dent in the form of food imports, especially staple food in the future.

4.2.4.3 What Is Next?

The first objective right now is to intensify and finish the process of humanitarian relief and rescue, provide food, transport the impacted populations to safety, and provide them with essentials like food, shelter, and urgent medical treatment. Primary care services and preventative healthcare methods must be reinstated at the same time, especially for women and children. Pregnancy care and facilities for safe deliveries should be prioritized considering the disruption of immunization and health outreach activities. Representatives of the civil society and nongovernmental organizations in Pakistan are driving a lot of the current response, which includes a national conversation and agreement on water conservation, interprovincial discussion on water distribution and new reservoirs, and renovation of the insufficient and dysfunctional drainage systems. Table 4.4 presents some current and future scenarios of rainfall and temperature for Pakistan in order to develop some resilience plans ahead. Without a national process of healing and political rapprochement, none of these imperative acts will be achievable. It is certainly legitimate to seek climate punishment to rebuild for a nation that generates less than 05% of global carbon emissions each year (Worldometer 2022). Infrastructure and flood control measures in Pakistan (Sachs 2022).

This case is grounded on systematic review of climate change impacts on agriculture sector of Pakistan in relation to floods and food insecurity. Climate change has initial indications like long-term decreases in available freshwater from glaciers and contamination of water supplies, reduction in crop yields, increased incidence of heat-related illness and death, water borne (cholera and diarrhea) and vector borne (malaria and tick borne) diseases, malnutrition, and poor hygiene from water scarcity, floods and droughts. Therefore, the adaptation techniques should be taken to mitigate this climate change. One key issue requiring prompt consideration in horticulture is the water circle of Pakistan, particularly the accessibility of water for water system would be an important perspective for deciding the future effects of environmental variations on Pakistan's crop yields. It is especially imperative that a creating

Table 4.4 Impacts of climate change on Pakistan and its effects on agriculture (current and future scenarios).

Country	Temperature	Rainfall	References
Pakistan	Predicted change in temperature ranges from 2.33 to 4.78°C along withphotocopying CO_2 concentration	Thwarts in rainfall frequency occasions	Din et al. (2022)
Pakistan	Increase in mean temperature on per annum basis from 1.3 to 1.7°C	Increment in the rainfall in rest of the season (Kharif) by 6–13%	Khan et al. (2022)
Pakistan	Winter temperature increment by 1–4°C plusincrement in CO_2 concentration	Almost 20% hike in precipitation and rainfall in summers (Kharif growing season)	Nadeem et al. (2022)
Pakistan	**Prediction or forecasting** *Mean temperature °C increment* **Year 2020** Kharif season: 0.87–1.12 Rabi season: 1.08–1.54 **Year 2050** Kharif: 1.81–2.37 Rabi: 2.54–3.18 **Year 2080** Kharif: 2.91–4.62 Rabi: 4.14–6.31	**Prediction or forecasting** *Variation in rainfall (%)* **Year 2020** Kharif: 1.8–5.1 Rabi: −2.0 to 4.7 **Year 2050** Kharif: 7.2–10.5 Rabi: −9.2 to 3.8 **Year 2080** Kharif: 10.1–15.2 Rabi: −24.8 to 4.5	Ansari and Grossi (2022)

nation like Pakistan will be most presumably more appropriate for adjustment to environmental change than alleviation. Mindfulness for change in environment and its unfavorable impacts for the jobs of the populace must be elevated, and existing versatile processes must be utilized to most extreme impact, for example, on account of rural credits.

4.2.4.4 Concluding Remarks

The Pakistani government, with the army's assistance, began extensive rescue efforts. Foreign humanitarian groups dispatched air personnel to Pakistan to assist with the crisis, including those from Turkey, the

United States, England, Japan, Saudi Arabia, and many other nations. While catastrophe victims and the media condemned the United Nations, international NGOs, local NGOs, and philanthropic groups, they claimed to have done a lot of help (Amen et al. 2020). The government's relief efforts were criticized by the media, who also criticized lawmakers for their lack of seriousness, which led to poor administration of the rescuers (Britannica 2022). There are significant distinctions from the previous flood tragedy in Pakistan in 2010 (Bhutta and Bhutta 2010). The National Disaster Management Authority and its provincial counterparts failed to act quickly to reinforce embankments, clear key drains, or store supplies for a probable mass emigration. With more than 1460 primary health institutions affected and 432 were destroyed, the COVID-19 epidemic and the economic crisis have further crippled Pakistan's already fragile health sector. Delivery of food commodities, medical treatment, availability to necessary medications, regular immunizations, and disease surveillance have all been hampered in areas impacted by flooding. The legislature of Pakistan needs to set up plans that can remunerate ranchers for their misfortunes and yield impacts as a result of actualizing certain versatile measures. Adaptation techniques should be done to eliminate the risk of climate change in livestock and aqua culture production. Forests are declining day by day due to anthropogenic activities, so more plantations should be done to overcome this problem as "Go green live green."

4.2.5 Climate Change and Vulnerability to Agriculture System: Case of Thailand

4.2.5.1 Background

Thailand is in middle of Southeast Asia and once known as Asian Tiger, covering an area of about 513,115 square kilometers. Climate of the country is tropical or savannah, with the average 18–34°C temperatures and an average rainfall of over 1500 mm/annum. Thailand has a biodiversity of plants and animals and is the foundation of production in the agricultural sector. Thailand has a mixed economy of UNDP[5]-provided countries andis "the highly successful" country in East Asia. The number of citizens moving from poverty into the middle class is increasing. Export value is quite high for Thailand in the world. For this reason, Thailand is so diverse and has intensive land use. Thailand has a plenty of low plain land and rain volume as well as water basin which are appropriate for cultivation. These

5 United Nations Development Program.

natural conditions bring a number of benefits to the country by creating the agricultural product providing in domestic and international level. While nowadays, the global warming issue brings the problems of climate change around the world affecting the natural flow, for example, relatively high temperature, water preservation, soil fertility as well as increase in precipitation which causes flooding to impact directly on agricultural system. World Bank's estimations showed economic damages and losses such as 1425 trillion baht (US$46.5 billion) because of flooding. Most of this was related to Thailand agricultural products.

4.2.5.2 Relationship Between Climate Change and Floods in Thailand

La Niña phenomenon[6] resulting from climate change caused the big problems in Thailand in 2011. Five tropical depressions attacked Thailand and increased the volume of precipitation. The capacity of the dams cannot support the huge volume of water and difficult to manage, so the mega flood finally began. According to the National Oceanic and Atmospheric Administration (NOAA), in 2011, the Oceanic Nino index which is the indicator of the severe of climate illustrates minus value (Figure 4.13). It means the happening of La Niña phenomenon in severe levels caused the change of warm current direction in Pacific Ocean. The warm current flowed into Southeast Asia region (Figure 4.14) caused the temperature of sea water to dramatically increase. High evaporation created much of rain cloud and finally became the tropical depressions, including Haima, Nok-Ten, Haitang, Nesat, and Nalgae. The tropical depression caused the volume of precipitation to increase extremely which was highest in 23 years history. The volume of rainwater of whole country calculating from January to 4 November 2011 was 1856 mm (Figure 4.15). Moreover, we can see that in 1996, 1999, 2000, 2008, and 2011 which had the La Niña phenomenon, the precipitation in Thailand also increased; therefore, climate change has the significance in mega flood in 2011.

4.2.5.3 Flood Impacts on Agriculture of the Country

The mega-flood in 2011 impacted the economy of Thailand thoroughly in terms of supply, including agriculture, industry, and service. In contrast, it also had been affected the economic demand around 190 thousand million

6 "El Niño and La Niña are opposite phases of what is known as the El Niño-Southern Oscillation (ENSO) cycle. The ENSO cycle is a scientific term that describes the fluctuations in temperature between the ocean and atmosphere in the east-central Equatorial Pacific (approximately between the International Date Line and 120° West)."

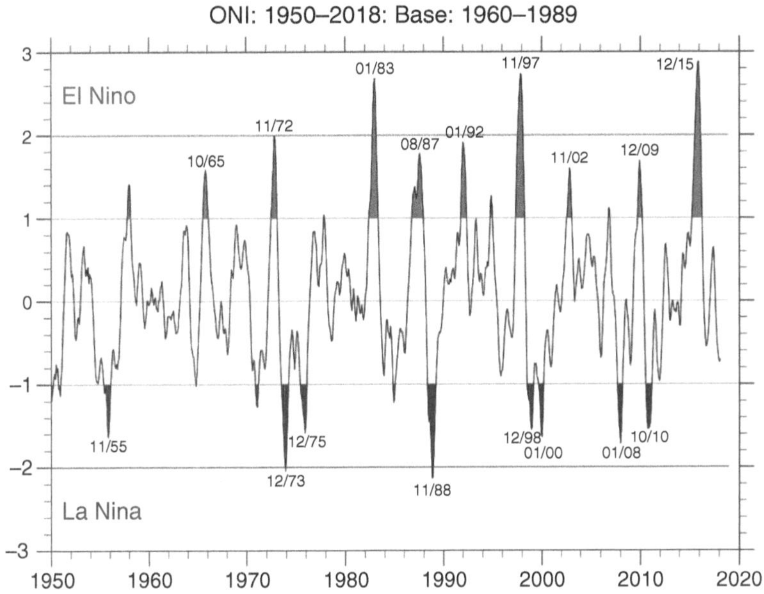

Figure 4.13 Oceanic Nino (ONI) Index. NCEI, 2011.

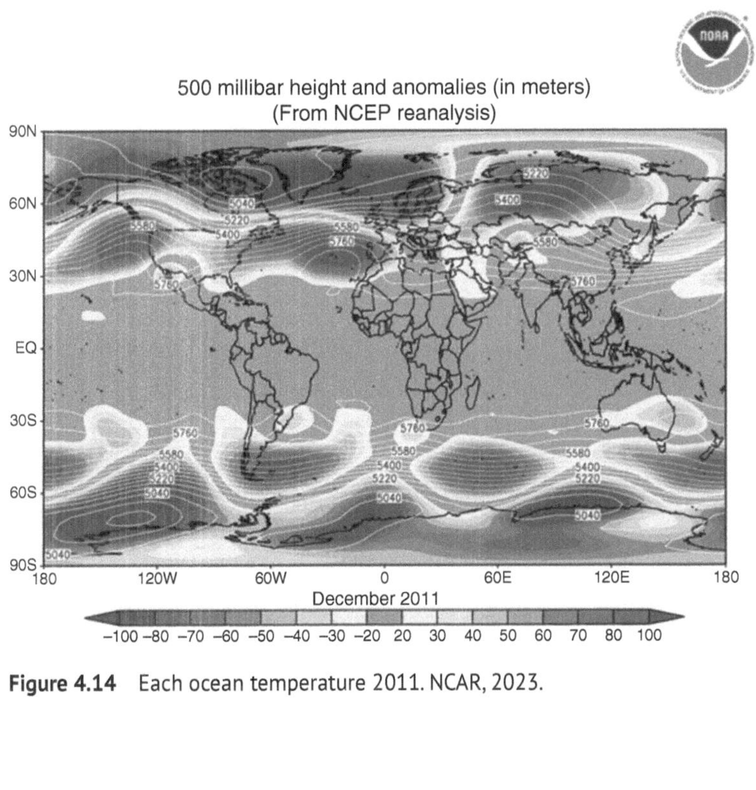

Figure 4.14 Each ocean temperature 2011. NCAR, 2023.

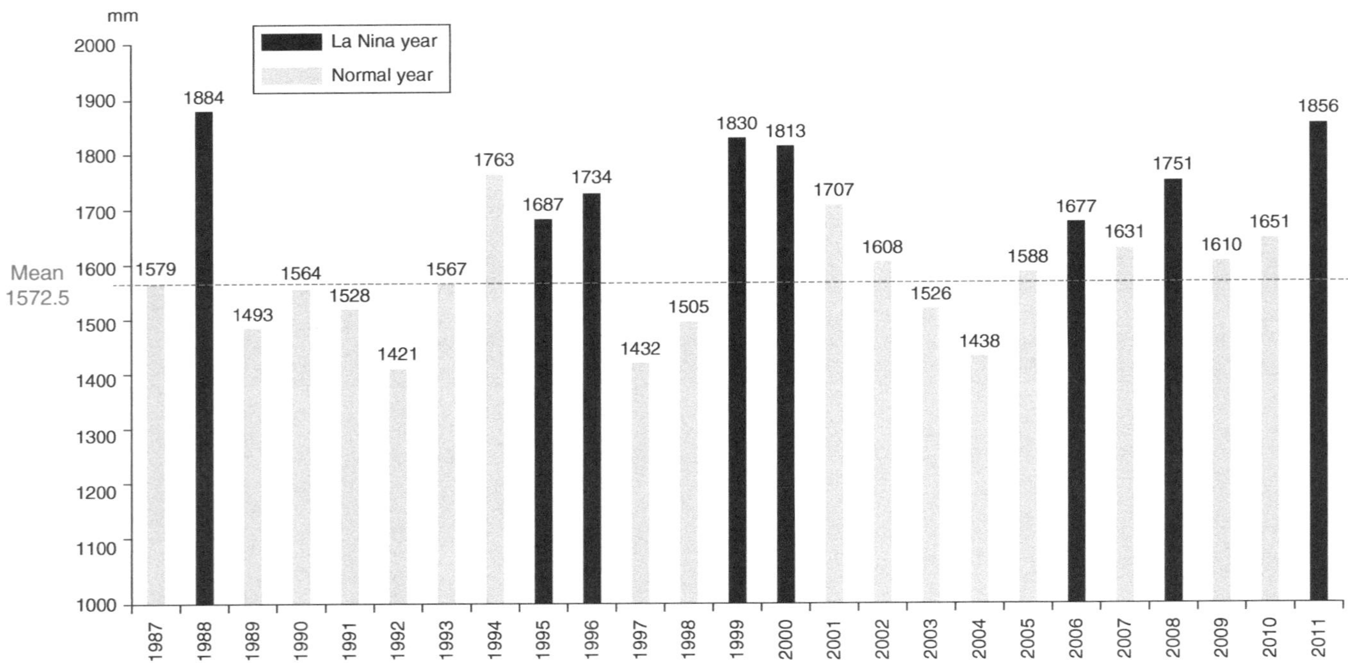

Figure 4.15 Average rainfall of Thailand 1987–2011. *Source:* Adapted from Office of Agricultural Economics, Bangkok, 2012.

baht.[7] It means economic had been declined 1.81%. Furthermore, the lack of consumer product was the cause of inflation (Kabiri 2016).

Agriculture was impacted very much, 17,578 sq.km of agricultural area was damaged. In addition, 194,012 fish cages were affected, in total 13,282,622 aquatic animals (Table 4.5). While the office of agricultural economics showed that 6 million paddies were also damaged, which caused the rice product volume in 2011–2012 to decrease to 19 million tons from the predicted number of 25 million tons. The impact was not only on agriculture itself but also related to national economic. Global supply chain of agriculture products of Thailand was hugely effected and interrupted by these flooding issues in that particular period and afterward (Wijitkosum 2016).

4.2.5.4 Flood Victims and Relevant Public Sector (Responses)

From standpoint of the agriculturists who had planted the second rice in Chum Saeng locale at Nakhon Sawan region. They showed that the impact of unseemly administration and coordination deterred surge water stream.

Likewise, they additionally pointed out about an official cautioning educated past the point of no return and was much disarray there. Harm was

Table 4.5 Summary of flood impact damages on agriculture of Thailand 2011.

Category	Unit	Damages (million bath)
Total damaged area	17,578 sq.km	–
Paddy field	14,062 sq.km	22,247
Cassava field	3,516 sq.km	9,393
Fishery	194,012 cages	2,910
Live stock	13,282,622	–
Chicken	10,626,098	80
Cattle	664,131	1,494
Pig	1,992,393	2,989
Total	–	39,113

Source: Adapted from Office of Agricultural Economics, Bangkok, 2012.

7 Thai currency. 1 USD = 34 baht as of 17.10.2022.

caused by the planted region of second rice, and everything of them included rice items were not secured to take care of any abrupt issue by vast surge. As per the specialists of the Thai Meteorological Department, Royal Irrigation Department, and Department of Disaster Prevention and Mitigation (Wijitkosum 2016), they acknowledged the improper administration. What's more, there was no facilitated and coordinated effort between different organizations, bringing about working copy or duplication in numerous zones of the CPRB. Like the heads in all areas in a fiasco region additionally alluded that there was an official cautioning came past the point of no return, and indistinct adapting and counteractive action designs from the focal government region.

4.2.5.5 Concluding Remarks

Climate change at the global level can affect agricultural system in region and local level by the impacts from the change in natural aspects. Especially, the increase in rain volume which causes flood disaster directly affects the agriculture. The damages thoroughly have impact on the economic and on human well-being. Therefore, the government and relevant institution should develop mitigation and adaptation policy. Right off the bat, the pertinent nearby associations in ponder territory ought to advance the cataclysmic event readiness and methods for dealing with stress arranging from climatic change for lessening of both immediate and roundabout impacts, for example, woods and upstream protection in the northern district, and make and work a neighborhood surge cautioning framework, and so on. Additionally, the engaging networks and ranchers in the administration of normal assets and rural creations offer learning and inquire about monetary help or crisis surge alleviation advances to cover harms in part of agribusiness area. Besides, people in general area associated with surge ought to have clear direction of the water administration and surge control extends as execution of aversion programs in long haul and commitment of the checking procedure from nongovernment organizations and make water administration activities to better plan for flooding along the CPRB.

4.3 Conclusions

This chapter was based on case studies from different developing countries of Asia. This is very much in line with the main theme of this book. It was tried to present concise yet significant case studies that show a robust nexus between climate change and its impacts on agriculture.

Key lessons that could be drawn out of this whole exercise are as follows: climate change has been everywhere in developing countries, even China is an emerging economy, and climate change has impacted in some parts in the form of floods, hurricanes, crop losses, and natural hazards. India and Pakistan are usually hit by monsoon rains almost every year, and Pakistan has been hit severely in the last 12 years as compared to India. Crop losses in 2022 are far more than 2010 flood disaster in Pakistan and that is/will directly impact the food availability/accessibility. Thailand has had a mega flood disaster in 2011, and since then there is no such a huge flood witnessed. Moreover, Thailand has planned and worked more toward resilience considering the vulnerable areas. Almost same is the case for Myanmar, that is, it is not severely hit by flood and did not face food shortages due to agriculture crop losses, though climate has impacted in that country as well. They are more toward building coastal resilience communities, working in the light of SDGs, and trying to develop resilient communities as much as better they can do. China and India have more dams and reservoirs and water management system than other countries studied in this chapter. They do face flooding and agriculture crop damages issues, but they have developed relevant resilience policies and plans and implement them effectively with the each passing day. Contrary to that, Pakistan has not learnt any lessons in these 15 years to cope with such challenges which may occur more frequently in the future. Resilience to climate change and adaptation of relevant action plans are damn necessary to cope. If same situation persists in the following years, then damages could be more than now. Therefore, coping strategies such as adaptation to climate change, building dams, creating awareness, preserving water, harvesting rainwater, and building resilient communities are some of the ways forward. Food security is directly linked to the agricultural crops, and those are most vulnerable to climatic changes happening rapidly in developing countries. Therefore, in order to develop food security measures, climate change-related adaptations and resilience need to be taken care of on priority basis by each sector of the economy.

Acknowledgment

I would like to acknowledge all my colleagues, reviewer(s), and Ph.D. students who helped me in revising and reviewing this chapter.

References

Adaptation to Climate Change (2017). https://www.google.com.sg/search?q=effect+of+climate+change+on+agriculture+in+Asia+images&biw=1366&bih=638&tbm=isch&tbo=u&source=univ&sa=X&ved=0ahUKEwibysLegJjUAhXKr48KHSx7CgAQsAQIIA#imgrc=XS4jSytIoYb1qM (accessed 30 September 2022).

Affoh, R., Zheng, H., Dangui, K., and Dissani, B.M. (2022). The impact of climate variability and change on food security in sub-saharan africa: perspective from panel data analysis. *Sustainability* 14 (2): 759.

Ali, A., Khan, T.A., and Ahmad, S. (2018). Analysis of climate data of Khyber Pakhtunkhwa, Pakistan. *International Research Journal of Engineering and Technology* 5 (5): 4266–4282.

Ali, G., Murtaza, M., Bashir, M.K., and Abbas, S. (2021). Drinking-water efficiency, cost of illness, and peri-urban society: an economic household analysis. *PLoS One* 16: e0257509. https://doi.org/10.1371/journal.pone.0257509.

Al Jazeera Media(2022). Floods in Pakistan.https://www.google.com.hk/search?q=Flood+disaster+map+of+Pakistan+2022&newwindow=1&source=lnms&tbm=isch&sa=X&ved=2ahUKEwjX6N7F4uf6AhUErmoFHUM-Au4Q_AUoAXoECAEQAw&biw=1440&bih=674&dpr=2#imgrc=ooe4-d-uWkGekM(accessed 16October 2022).

Amen, H., Mahmood, S.A., Batool, S. et al. (2020). River profile modeling through surface deformation using RS/GIS, acase study Swat River. *International Journal of Innovations in Science & Technology* 2 (3): 75–88.

Ansari, R. and Grossi, G. (2022). Spatio-temporal evolution of wet–dry event features and their transition across the Upper Jhelum Basin (UJB) in South Asia. *Natural Hazards and Earth System Sciences* 22 (2): 287–302.

Rahman, A. (2010). *Disaster Risk Management: Flood Perspective*. Germany: VDM Verlag.

Aung, G.R. (2022). *Infrapolitics: The Political Life of Infrastructure in a Myanmar Economic Zone*. Columbia University.

Bandara, J.S.A. and Cai, Y. (2014). The impact of climate change on food crop productivity, food prices and food security in South Asia. *Economic Analysis and Policy* 44: 451–465.

Bashir, M.K. and Schilizzi, S. (2013). Determinants of rural household food security: a comparative analysis of African and Asian studies. *The Journal of the Science of Food and Agriculture* 93 (6): 1251–1258.

Bhutta, Z.A. and Bhutta, S.Z. (2010). The unfolding human tragedy in Pakistan: fighting alone. *Lancet (London, England)* 376 (9742): 664–665.

Boretti, A. and Rosa, L. (2019). Reassessing the projections of the world water development report. *NPJ Clean Water* 2 (1): 1–6.

Britannica, T. (2022). Pakistan floods of 2010. In: *Encyclopedia Britannica*. Britannica.

Crippa, M., Solazzo, E., Guizzardi, D. et al. (2021). Food systems are responsible for a third of global anthropogenic GHG emissions. *Nature Food* 2 (3): 198–209.

Din, M.S.U., Mubeen, M., Hussain, S. et al. (2022). World nations priorities on climate change and food security. In: *Building Climate Resilience in Agriculture* (ed. W.N. Jatoi, M. Mubeen, A. Ahmad, et al.), 365–384. Springer.

Ding, Q., Chen, X., Hilborn, R., et al. (2017). Vulnerability to impacts of climate change on marine fisheries and food security. *Mar. Policy* 83: 55–61. https://doi.org/10.1016/ j.marpol.2017.05.011.

Fanzo, J., Rudie, C., Sigman, I. et al. (2022). Sustainable food systems and nutrition in the 21st century: a report from the 22nd annual Harvard Nutrition Obesity Symposium. *The American Journal of Clinical Nutrition* 115 (1): 18–33.

FAO (2022). Monsoon-flood situation update-Pakistan. https://www.google.com.hk/search?q=Flood+disaster+map+of+Pakistan+2022&newwindow=1&source=lnms&tbm=isch&sa=X&ved=2ahUKEwjX6N7F4uf6AhUErmoFHUM-Au4Q_AUoAXoECAEQAw&biw=1440&bih=674&dpr=2#imgrc=sUVJ-xznk2V2PM&imgdii=AQhumpe4Tzy2FM (accessed 15October 2022).

Gostin, L.O. (2021). *Global Health Security: A Blueprint for the Future*. Harvard University Press.

Han, X., Hua, E., Engel, B.A. et al. (2022). Understanding implications of climate change and socio-economic development for the water-energy-food nexus: a meta-regression analysis. *Agricultural Water Management* 269: 107693.

Hoegh-Guldberg, O., Jacob, D., Bindi, M. et al. (2018). Impacts of 1.5 C global warming on natural and human systems. *Global Warming of 1.5° C* 3: 175–311.

Hong, C., Burney, J.A., Pongratz, J. et al. (2021). Global and regional drivers of land-use emissions in 1961–2017. *Nature* 589 (7843): 554–561.

Impacts of Climate Change(2017). Effect of climate change on agriculture in Asia. https://www.google.com.sg/search?q=effect+of+climate+change+on+agriculture+in+Asia+images&biw=1366&bih=638&tbm=isch&tbo=u&source=univ&sa=X&ved=0ahUKEwibysLegJjUAhXKr48KHSx7CgAQsAQIIA#imgrc=Lgs_HgPglWmGnM(accessed 30 May 2017).

IPCC (2013). Intergovernmental Panel on Climate Change. Climate Change 2013: The Physical Science Basis. Final Draft Report of Working Group I, Stockholm, Sweden.

IPCC WG II (2014). Climate change 2014: impacts, adaptation, and vulnerability. Chapter 22 (Africa).

Jain, M.S. (2021). A mini review on generation, handling, and initiatives to tackle construction and demolition waste in India. *Environmental Technology & Innovation* 22: 101490.

Kabiri, N. (2016). Public participation, land use and climate change governance in Thailand. *Land use policy* 52: 511–517.

Khan, Z., Khan, F.A., Khan, A.U. et al. (2022). Climate-streamflow relationship and consequences of its instability in large rivers of Pakistan: an elasticity perspective. *Water* 14 (13): 2033.

Kohnert, D. (2021). On the socio-economic impact of pandemics in Africa-Lessons learned from COVID-19, Trypanosomiasis, HIV, Yellow Fever and Cholera. *Trypanosomiasis, HIV, Yellow Fever and Cholera* (6 May 2021). Hamburg: GIGA – Institute of African Affairs.

Yenneti, K., Tripathi, S., Wei, Y.D. et al. (2016). The truly disadvantaged? Assessing social vulnerability to climate change in urban India. *Habitat International* 56: 124–135.

Krishnan, M. and Lakra, W.S. (2013). Coastal and Marine Fisheries Management in. *Coastal and Marine Fisheries Management in SAARC Countries* 196: 187.

Leakey, R.R.B. (2018). Converting 'trade-offs' to 'trade-ons' for greatly enhanced food security in Africa: multiple environmental, economic and social benefits from 'socially modified crops. *Food Security* 10 (3): 505–524.

Li, D., Li, Z., Liu, Z., Yang, Y., Khoso, A. G., Wang, L., &Liu, D. (2023). Climate change simulations revealed potentially drastic shifts in insect community structure and crop yields in China's farmland. *Journal of Pest Science*, 96 55–69.

Lu, Y., Zhang, Y., Hong, Y. et al. (2022). Experiences and lessons from agri-food system transformation for sustainable food security: a review of China's practices. *Foods* 11 (2): 137.

MCCA (2018). Impact of Climate Change and the Case of Myanmar. Myanmar Climate Change Alliance. Copyright Myanmar Climate Change Alliance2018. https://myanmarccalliance.org/en/climate-change-basics/impact-of-climate-change-and-the-case-of-myanmar/ (accessed 09 October 2022).

Meyers, W.H. and Kalaitzandonakes, N. (2015). World population, food growth, and food security challenges. In: *Food Security in an Uncertain World*. Emerald Group Publishing Limited.

MoNREC (2016).Myanmar Climate Change Strategy and Action Plan, the Republic of the Union of Myanmar, Nay Pyi Taw, Myanmar.

Nadeem, M., Nazer Khan, M., Abbas, G. et al. (2022). Application of CSM-CANEGRO model for climate change impact assessment and adaptation for sugarcane in semi-arid environment of Southern Punjab, Pakistan. *International Journal of Plant Production* 16: 443–466. https://doi.org/10.1007/s42106-022-00192-6.

NCAR (2023). Actionable Earth System Science with and for Society. National Center for Atmospheric Research. https://ncar.ucar.edu/what-we-offer/education-outreach/2023-2025-cohort (accessed 12 December 2022).

NCE (2011). 2011 National Climate Report. National Centers for Environmental Information. https://www.ncei.noaa.gov/access/monitoring/monthly-report/national/2011 (accessed 22 October 2022).

Nhamo, L., Matchaya, G., Mabhaudhi, T. et al. (2019). Cereal production trends under climate change: impacts and adaptation strategies in southern Africa. *Agriculture* 9 (2): 30.

Noort, M.W.J., Renzetti, S., Linderhof, V. et al. (2022). Towards sustainable shifts to healthy diets and food security in sub-Saharan Africa with climate-resilient crops in bread-type products: afood system analysis. *Foods* 11 (2): 135.

Nugee, R. (2022). The operational advantages of a Greener defence: what should defence be prepared for in the next 20–30 years as a result of climate change? In: *Innovative Technologies and Renewed Policies for Achieving a Greener Defence*, NATO Science for Peace and Security Series C: Environmental Security (ed. G. Iacovino and M. Wigell), 17–33. Dordrecht, The Netherlands: Springer.

Office of Agricultural Economics (2012). Agriculture Economics Data. Office of Agricultural Economics, Bangkok, Thailand. https://www.oae.go.th (accessed 20 November 2022).

Orgeret, F., Thiebault, A., Kovacs, K.M. et al. (2022). Climate change impacts on seabirds and marine mammals: The importance of study duration, thermal tolerance and generation time. *Ecology Letters* 25 (1): 218–239.

Pakistan Economic Survey (2014). *Economic Survey of Pakistan, Agriculture*. Islamabad, Pakistan: Government of Pakistan.

Pakistan Economic Survey (2015). *Agriculture. Economic Survey of Pakistan, Agriculture*. Islamabad, Pakistan: Government of Pakistan.

Pandit, M.K. (2021). Climate change, urbanization and impact on natural environment: The Indian Scenario. In: *Urban growth and environmental issues in India*, 173–187. Springer.

Petry, N., Jallow, B., Sawo, Y. et al. (2019). Micronutrient deficiencies, nutritional status and the determinants of anemia in children 0–59 months of age and non-pregnant women of reproductive age in the Gambia. *Nutrients* 11 (10): 2275.

Piao, S., Ciais, P., Huang, Y. et al. (2010). The impacts of climate change on water resources and agriculture in China. *Nature* 467: 43–51 (2010).

Rehamn, S.(2015). COP 21 meeting in Paris and Pakistan. *World Tribune Pakistan*. http://www.worldtribune.com.pk (accessed 14 November2016).

Sabin, N.S., Calliope, A.S., Simpson, S.V. et al. (2020). Implications of human activities for (re) emerging infectious diseases, including COVID-19. *Journal of Physiological Anthropology* 39 (1): 1–12.

Sachs,J.D. (2022). Pakistan and the fight for climate justice. Project Synsdicate.https://www.project-syndicate.org/commentary/pakistanflooding-climate-justice-rich-countries-should-bear-cost-by-jeffrey-dsachs-2022-09 (accessed 13 September 2022).

Shahzad, N. and Amjad, M. (2022). Climate change and food security in Pakistan. In: *Sustainable Agriculture and Food Security*, 579–594. Springer.

Siddiqui, R., Samad, G., Nasir, M., and Jalil, H.H. (2012). *The Impact of Climate Change on Major Agricultural Crops: Evidence from Punjab, Pakistan*. The Pakistan Development.

Stuch, B., Alcamo, J., and Schaldach, R. (2021). Projected climate change impacts on mean and year-to-year variability of yield of key smallholder crops in Sub-Saharan Africa. *Climate and Development* 13 (3): 268–282.

Thein, H.M. (2015). *Development Planning and Climate Change in Myanmar*. Myanmar: Yangon.

Uhlenbrook, S., Yu, W., Schmitter, P., and Smith, D.M. (2022). Optimising the water we eat—rethinking policy to enhance productive and sustainable use of water in agri-food systems across scales. *The Lancet Planetary Health* 6 (1): e59–e65.

Wang, F. and Zhang, Z. (2022). Decoupling economic growth from energy consumption in top five energy consumer economies: a technological and urbanization perspective. *Journal of Cleaner Production* 357: 131890.

WHO (2022). Pakistan floods situation report issue 1. https://reliefweb.int/report/pakistan/who-pakistan-floods-situation-report-issue-1-30-august-2022 (accessed 30 August 2022).

Wijitkosum, S. (2016). The impact of land use and spatial changes on desertification risk in degraded areas in Thailand. *Sustain. Environ. Res.* 26 (2): 84–92.

World Health Organization (2018). *Managing epidemics: key facts about major deadly diseases*. World Health Organization.

Worldometer (2022). Pakistan' CO2 emissions. https://www.worldometers.info/co2-emissions/pakistan-co2-emissions(accessed 16 September 2022).

Xie, S., Ding, W., Ye, W., and Deng, Z. (2022). Agro-pastoralists' perception of climate change and adaptation in the Qilian Mountains of northwest China. *Scientific Reports* 12 (1): 1–15.

Xu, X., Sharma, P., Shu, S. et al. (2021). Global greenhouse gas emissions from animal-based foods are twice those of plant-based foods. *Nature Food* 2 (9): 724–732.

5

Agricultural Revolutions and Food Security

Tahira Sadaf, Sana Sehar, and Sarfraz Hassan

Institute of Agricultural and Resource Economics, University of Agriculture, Faisalabad, Pakistan

5.1 Background

The economies of developing countries rely heavily on agriculture. Ensuring food security is a big challenge for such agro-based developing countries. Agricultural revolutions like "the Green Revolution" play a crucial role in fighting poverty and bringing food security. Agriculture in these countries is mostly based on small holdings, and it is predominantly meant for a household's subsistence. Such setup demands for coordinated support from the government and private sector for a successful agricultural transition. To compete at the international market level and promote food self-sufficiency, these countries need to have required domestic reform and structural intervention. Developing countries need to improve indigenous practices along the whole value chain of existing agricultural commodities and diversifying agricultural products. The GR helped nations fight hunger through several integrated interventions like using high-yielding seed varieties, fertilizers/pesticides, and agro-mechanization. The Second Agricultural Revolution is in need of time in the form of a revolution emerging from the dairy sector of agriculture with the aim of reducing stunting and undernourishment among the population, thus increasing incomes, sustaining livelihoods, and alleviating poverty. Thus, food security plays a significant role in reducing poverty in developing nations.

In this chapter, the role of revolutions in enhancing food security in the history of agriculture is discussed. It presents a discussion on opportunities

Food Security in the Developing World, First Edition. Edited by Muhammad Khalid Bashir, Steven G.M. Schilizzi, and Ghaffar Ali.

and challenges associated with the White Revolution. The first section presents the background of agricultural revolutions, wherein the earliest "Neolithic Revolution" and the Second Agricultural Revolution are discussed in detail. The second section is about the Green Revolution. It presents the background of the GR, its benefits and drawbacks, and the post-GR situation. The third section is about the dairy subsector of agriculture's "the White Revolution." It is discussed how it took place in India and Brazil. Perspectives of the WR in Pakistan are discussed. The last section concludes this chapter.

5.2 Agricultural Revolutions: Historical Perspective

Agriculture is a source of food, shelter, and clothing for the human race. The importance of agriculture cannot be denied when we are concerned about food and nutrition security in developing countries. Food surplus produced in rural agricultural areas provides a base for industrialization in the urban part. Any urban development would have been impossible without the support from agriculture. Globally, agriculture has remained the most important enterprise since early human civilization. Agriculture is the source of 27% employment globally in 2019. The share of agriculture in total world employment in the case of low-income countries is 62% according to the World Bank in 2021. In South Asia and sub-Saharan Africa, 45% and 61% of employment are generated by agriculture, respectively, during 2021. With the pace of development of human civilization, agriculture has also developed from time to time. History reveals that there have been several revolutionary changes in agriculture over time. Although this change was not the same geographically, the transformation from agriculture as a source of subsistence to commercial activity was a major change observed over time.

The first-ever agricultural revolution is famous for the name "the Neolithic Revolution." It was a process of transition from hunting and gathering food items to the domestication of a variety of valuable plants and animals using different agricultural practices, resulting in increased food production for daily use. This transition happened at a different pace in different parts of the world. Nevertheless, majority of masses of this era had dietary homogeneity given their substantial reliance on a limited range of food items. Domestication of animals for getting milk and meat increased dietary diversity and food security. Reportedly, the majority of world's population underwent this transition during the period between 10,500 and 6,000 years ago. The Neolithic Revolution provided the basis for the subsequent transitions and revolutions in human civilization. It involved the domestication of different useful plants and animals for the production of food for daily use and as surpluses. The first crops domesticated by the Neolithic farming communities included wheat and barley along with lentils, chickpeas, peas, and flax. Later on people from Asia started growing rice and millet. In the Mexican region, squash and maize were also cultivated. Plowing of land started for the first time in the human history. The population increased, and people started outmigration toward cities for having better resources. This enabled human beings to develop permanent settlements in the form of cities and rural areas with improved social structures. The Neolithic Revolution, however, also had some drawbacks as well. It resulted in the emergence of new diseases carried by the domestication of plants, animals, and cultural exchange among different human communities. Initiation of using weapons appeared to be a threat to many species of animals and humans. The Neolithic Revolution was also the start of environmental changes and ecological crisis. Neolithic humans domesticated animals predominantly for getting milk and meat, which added some variety to their diet. The first farm animals included were sheep and cattle. Buffalo and yak were also domesticated in the Asian region. Draft animals like ox, donkey, and camel were used for transportation later during 4000 BCE.

The Second Agricultural Revolution, "the British Agricultural Revolution," happened during the mid-seventeenth century. It was rooted in Western Europe, and within two centuries it spreaded to North America. It was marked by an extraordinary increase in agricultural production because of increases in land and labor productivity, crop rotation, and improved transportation facilities. Technological advancement took a new pace which already was initiated by the ancient Romans, and then later, in the European Middle Ages, North-Western Europe invented heavy plow as well as modern yoke, which did not have animals choke at the neck. Then in the 1700s, France rationalized crop rotations as a way to reduce fallow. So, improvements in agricultural technology have been happening steadily over time, and in other parts of the world like China, there were also gradual improvements in agricultural practices and mechanization. In addition to agricultural revolutions, there has also been a steady agricultural technological evolution. Commercialization and trade of surplus product(s) and urbanization were prevalent. Commercialization practices supplemented subsistence farming. People started using tools and equipment with great land and labor productivity. Transportation facilities were modernized through technology that could link farms to urban markets. All such innovations increased crop and livestock productivity. This revolution was characterized by capital–labor substitution and the use of man-made fertilizers and herbicides, etc., and the creation of positive feedback loops where one progress allowed another which further allowed another, all feeding on themselves. This created a "progress dynamics." This was the first technological revolution with the emergence of industries. The rise in productivity resulted in a decreased agricultural share of the labor and increased urbanization resulting in increased industrialization. Improved productivity resulted in an excess of labor force, which was ultimately relieved from agriculture to industry leading to the Industrial Revolution. The Second Agricultural Revolution has hence been reported as a trigger for the Industrial Revolution. The Industrial Revolution also smoothened the way for mega food chains and corporations resulting in decreased prices of food and an overall increase in food accessibility and utilization. In addition to the agricultural revolutions, there has also been a steady agricultural technological evolution. The Industrial Revolution remained a success; however, a staggering proportion of global population was still hungry. This represents the famous Malthusian dynamics, where food per person does not increase over time because of an exponential increase in population that outpaces food production. The situation also called for another revolution for bringing food security.

5.3 The Green Revolution

Agricultural revolutions discussed earlier were part of how human civilization evolved over time. The Green Revolution was the third Agricultural Revolution after the Neolithic Revolution and the British Agricultural Revolution taking place between 1950 and late 1960s. It was somehow a more planned set of agricultural technology transfer initiatives, plant breeding, and genetic engineering that enhanced agricultural productivity in different parts of the globe. It featured the use of hybrid and highly productive varieties, particularly dwarf varieties of cereals, the use of synthetic fertilizer/agrochemicals, advanced methods of irrigation, agro-mechanization, and advanced cultivation practices. The GR led to increased area under crops and high productivity of crops, particularly cereal crops.

5.3.1 The Green Revolution: Background

The term "Green Revolution (GR)" was first used by Norman Borlaug, who was an American Biologist. According to him, the GR means rapid increase in crop yields and land productivity. The GR is a mega revolution in the field of agriculture that has played an important role in fighting hunger. The GR roots back to Mexico, where in the 1940s, the US Vice President being worried about the increasing hungry population in Mexico met with Raymond Fosdic, who was the president of the Rockefeller Foundation (Kohler 2007). As a result of their meeting, the Mexican government established the Mexican Agricultural Program (MAP), in mutual collaboration with the Roymond's Rockefeller Foundation. The ultimate goal of this program was developing the Mexican agricultural sector, with the help of innovative technologies. The team working for the program led by J. George Harrar included Dr. Norman Borlaug, and others followed a three-step strategy in order to improve the yields of Mexico's basic cereal crops. First, the introduction of high-yielding varieties (HYVs) of food crops through advanced technologies and training local agronomists and other agriculture professionals (2007). Wheat rust, a fungal disease, was a challenge for Mexican agricultural scientist at that time. Hence, Borlaug with other agricultural professionals developed a rust-resistant variety of wheat (Ganzel Group 2007). After that, Borlaug focused on ways to enhance the yield of wheat. He developed a high-yielding dwarf variety of wheat that was also resistant to diseases. In the short period of 1956–1971, Mexico became self-sufficient in wheat production (Ganzel Group 2007). With the passage of time, growth in population resulted in increased demand for food in Mexico, which compelled it to import wheat. This represents the famous Malthusian dynamics, where food per person does not increase over time because of an exponential increase in population that outpaces food production. The Program team continued its efforts of introducing decease-resistant HYVs of crops with increased use of fertilizers and pesticides. Such varieties were capable of growing faster with the help of adequate usage of irrigation, fertilizers, and pesticides in contrast to the traditional varieties (Hicks 2011). Norman Borlaug was awarded the 1970s Nobel Peace Prize for introducing the "miracle seeds" (Shiva 1991).

5.3.2 Benefits of the Green Revolution

The GR brought a package of immense changes in conventional farming society. The developing countries became successful in changing their status to net exporters from net importers of grains. A good example here is Vietnam: from 1980 to 2000 and beyond, it totally changed its status on import–export of rice, even though its population increased. As a result of the GR, such countries improved their per capita food consumption besides improving the nutritional status of their population, particularly children. The results of the GR technologies became famous and were later adopted by the developing countries of South Asia, Latin America, and Africa in order to feed their growing population. Efforts to meet the requirement of food for the increasing population within South Asia were started ahead of the GR era in the 1950s, where the production of food grain crops in Pakistan increased because of dedicating increased agricultural land to the production of crops from 1950s to 1960s, as well due to increase in agriculture productivity. Similarly, Sri Lanka and Bangladesh experienced an increase in crop yields, in the "pre Green Revolution" period (Farmer 1981).

5.3.2.1 Increase in Production and Export of Food Cereals

The introduction of dwarf varieties of rice and wheat during the late 1960s increased the crop yields. In India, wheat production increased to three types in 1971 from 1964s yield of 9.9 million tons (Farmer 1981). Similar results were evident in Pakistan also. In the case of rice production, Pakistan and Sri Lanka met tremendous success, where the production of rice increased by 80% and 50%, respectively (Farmer 1981). During the GR period, the investment in irrigation systems increased as it was a prerequisite for producing high-yielding crops. Spending on the agricultural sector was increased. New credit schemes were introduced, and the government provided subsidies on inputs. Such interventions under the GR resulted in surplus production by developing countries most markedly Mexico, India, and Pakistan (IFPR India), and this created exporting opportunities. The Green Revolution has, however, introduced monoculture techniques and excluded many developing countries like the continent of Africa from the revolution.

5.3.2.2 Poverty Reduction

The GR was initiated with the motivation of bringing food security and reducing rural poverty. The GR–poverty alleviation link somehow remained complex. Although key beneficiaries were the progressive farms, nonetheless, the GR benefited poor villagers in many ways. It provided them with opportunities to enhance their own production, employment opportunities, and upgraded wages. Food surplus resulting from the GR in adapting regions resulted in lower food prices that benefited poor people. They became able to afford nutritious food through the cultivation of high-yielding crops. Resultantly, per capita food supply increased and requirement of masses for daily protein and calories were successfully met (Hazell 2009). Poverty head count was reduced by 28% from 1975 to 1995 in developing Asia. In the case of South Asia, for instance, in India alone, poverty head count was reduced by 9% (Hazell 2009). This achievement would have been sustainable if the population rate of the benefiting countries of the GR had an appropriate growth rate of population.

5.3.2.3 Rise in Public Investments

Public investment rose in developing rural infrastructure, providing input subsidies, and introducing better credit schemes. The research and development boosted yields resulting in greater yield and greater export of developing countries increasing their national income. It was also accompanied by employment generation and improvement of local infrastructure particularly in the rural parts of these countries.

5.3.3 Drawbacks of the Green Revolution

During the Green Revolution, hybridization of plants played a major role, where dwarf varieties were introduced. Other attributes introduced in plants, particularly in cereals, included drought resistance, shorter growing season, and great responsiveness to the use of fertilizer, etc. Such features introduced in cereals resulted in many times yields, shorter season of plant growth, increased income of farmers, diversity in diet options for rural dwellers, upgradation of local infrastructure, and employment generation. There is no doubt that it made more food available than would have existed without it. Besides bringing such positive changes, the GR was, however, accompanied by some disadvantages.

5.3.3.1 Challenge of Reduction of Rural Poverty and Income Inequality

During the pre-GR era, increasing cropped area was a strategy to increase cereal production, which was replaced by cultivation of HYVs of crops during the GR period. Farmers welcomed this strategy for bringing quick and impressive results. The drawback with this technology was the overuse of chemicals in the form of fertilizers and pesticides. Although it reduced poverty on average and lifted socio-economic conditions of developing countries, however, the evidences reveal that the GR remained quite a failure in rural poverty reduction and sustainable utilization of the environment (Setboonsarng 2006). The impressive results achieved from the GR were not sustainable in terms of bringing income equality. Still, a difference of opinion prevails among different authors on the complex nature of the relationship between poverty reduction and the GR (Hazell 2009). However, it was evident from the available literature that the fruit of the GR could not reach the poorest farmers in rural areas because of a lack of access to advanced technology. In contrast, urban farmers could better use such facilities to benefit from the GR. Poor farmers were in trouble with reducing prices of produce, increasing prices of inputs, and hiking land rents and even landlessness (Andersen et al. 1985). Particularly in South

Asia, rain-fed agriculture was a more common practice; however, cultivating HYVs of crops demanded proper irrigation systems, which were lacking during the era of GR (Pingali 2012a). The facility of credit for the purchase of chemicals like fertilizers and pesticides was also comparatively more accessible to large farmers in contrast with small farmers (Feder et al., 1985). In other words, poverty in rural areas remained a challenge, while large-scale producers of urban areas gained from the benefits of the Green Revolution. Thus, benefits from the GR were not evenly distributed, which increased the poverty gap and inequality between the rural and urban areas of the developing countries. Although the GR reduced poverty on average and lifted up the socio-economic conditions of developing countries, however, evidence reveal that the GR remained quite a failure in rural poverty reduction and sustainable utilization of the environment (Setboonsarng 2006).

5.3.3.2 Gender Inequality

At the household level in agriculture, there exists discriminating gender roles and access to innovations. Women play a vital role in agricultural activities; nonetheless, they have limited access to and control over productive assets like land, labor, credit, and capital. The GR caused a reduction of employment opportunities, especially for women due to the advent of technology. Mechanization replaced traditional manual work such as post-harvest practices. The GR resulted in inadequate investment in research on crops, where need for gender-sensitive technology required for women that could reduce their workload was ignored (McIntyre et al. 2009). On the positive side, the GR improved average conditions, but on the negative side, it increased various types of inequalities.

5.3.3.3 Environmental Hazards

Farmers readily adopted the technologies and strategy of the GR for bringing quick and impressive results. Since such interventions involved intensive use of inputs, particularly fertilizer and pesticides, for better production governments at that time encouraged the use of inputs by providing subsidies, and farmers of developing countries who were not educated followed these techniques blindly with excess use of agro-chemicals (Hazell 2009). Thus, the focus was limited to adopting HYVs, which were input-intensive, and proper planning for clarifying an upper limit on the use of inputs was lacking at that time (Evenson and Pingali 2007). As a result, the quality of the environment was compromised and land degradation, water pollution, and increased emission of greenhouse gases (GHGs) became evident during the GR and afterward. The GR remained quite a failure as regards the sustainable use of the environment (Setboonsarng 2006).

5.3.3.4 Land Degradation

Advantages brought by the GR had compromised land quality, particularly in the South Asian region, where more than one-third agricultural lands degraded to a severe extent (Flavin and Young 1993). Land degradation in the long-run results in decline of potential yields, decreased input efficiency, and decreased value of land and its use and land abandonment (Scherr and Satya 1996). Inefficient irrigation systems further deteriorated the situation by bringing salinity and water logging resulting in low crop yield (Postel 1989). In India's Haryana, GR practices resulted in annual elevation of the water table and salinity (Singh 2000). Overall, in India during the 1980s, 10.5 million ha of irrigated land suffered from salinity and waterlogging. Similar situation had to be faced by Pakistan, where more than one-fourth of total irrigated land was degraded by salinity during the 1990s (Ghassemi et al. 1995). The situation worsened with the passage of time, and almost 61% of agricultural land was degraded till the early 2000s (Shah and Arshad 2006).

5.3.3.5 Water Issues

Excessive irrigation practices may overuse both the groundwater and surface water sources, ultimately leading to water scarcity. When groundwater is excessively used in a way that water draft is higher than water recharge, groundwater table will fall, forming dark zones (Singh 2000). This causes drying out of aquifers and induces user conflicts for water. For example, in the early 1990s in Punjab, India, where tube wells were used for irrigation, groundwater table was negatively balanced in 50% of agricultural regions, and the rate of water draft was twice the rate of water recharge. Canals also cause problems since they alter natural water occurrence, and examples of flood damage in low-lying areas can be seen in Bangladesh (Fitzgerald-Moore and Parai 1996).

5.3.3.6 Other Issues

Other issues found as by-products of the GR included water pollution created by excessive use of chemicals, excessive pesticides, and fertilizers which served as a threat to wildlife and human health. Biodiversity loss also happened as a result of it due to focus on HYV. In this regard, Bangladesh suffered a lot where more than two-third acreage of traditionally grown wheat got displaced by HYV (Khan and Shah 2009). Still some other issues included emission of GHGs, increased cost of production, and nutritional insecurity.

5.3.4 Post-Green Revolution Situation

The Green Revolution played a remarkable role in meeting food security in a modernized way, ultimately increasing the national income of many countries in the South Asia. Such rise in incomes accompanied with urbanization resulted in abrupt changes in the pattern of national diets, as a consequence demand increased for diversified foods, particularly in the case of products of livestock and fruits and vegetables (Dorjee et al. 2003; Joshi et al. 2007). Though increase in national income does not usually correlate with higher income for the poorest farmers, who instead may suffer most from land degradation, etc., there were some significant consequences of the Green Revolution. It failed to include poor people into loop, and geographically many regions remained unattended, and it was further hampered by ever-increasing global population. Another upset in the GR was it failed to include all types of land while doing agricultural conservation, which ultimately led to environmental issues, like misuse and poor management of input resources, poor management and use of irrigation water, and loss of natural biodiversity previously found in rural areas and crops. Thus, the GR had short-term benefits but long-term costs.

The era after GR that can be called as the post-Green Revolution era roughly starts in the late 1980s (Chand and Raju 2009) or the early 1990s (Fischer and Cordova 1998; Gollin et al. 2005), and it has been characterized by sustainable natural resource management instead of merely increased productivity of food cereals. Focusing on the quality of high-value crops, livestock increased with the increased use of the ISO-9000 and ISO-14000 series of norms. Poverty eradication became the objective of agriculture-related policymaking, for instance, the UN started setting Sustainable Development Goals. Research and development (R&D) in the agricultural sector targeted environmental issues linked with agriculture, by keeping farmers in the loop. In the post-GR era, nations have improved their R&D priorities by inculcating environmental concerns, poverty reduction, and sustainable resource management into the system. Participatory engagements in rural areas became a priority in all agriculture-related interventions (Hazell 2009).

After the GR era, there were substantial changes in food production and consumption patterns. According to the Cambridge World History of Food, people of South Asia (for instance India) used millets and barley as their staple food (Kiple and Ornelas 2000). They transferred their staples into fodder (like millet) (Nelson et al. 2019), which was evident from figures given by the Food and Agriculture Organization (FAO) for the period of 1962–2017, where production of millets declined and production of rice increased (Food and Agricultural Organization 2019; Smith et al. 2019). In the post-GR era, nutritional security became more important than mere food security. As out of a number of failures of the GR, micronutrient malnutrition was significant one (Pingali 2012b), where the decreased price of staple food items discouraged the productivity of micronutrient-rich crops like vegetables and pulses (Gomez et al. 2017).

5.4 The White Revolution

"The White Revolution"[1] points to the success of "Milk and Milk Products," the "White" products of livestock subsector of agriculture. Dairy farming is considered as comparatively a profitable utilization of land resources among its alternate uses. It has the potential to provide quality food to a nation. In many respects, the dairy industry occupies a special position among other sectors of agriculture. Milk is being produced daily and provides a consistent return to farmers, especially those working on very small scale. As production of milk and milk products is quite labor-intensive activity, it provides employment to many persons. It would not be wrong to call milk as "white gold" because of a long value chain of quality goods associated with it. This precious commodity can be used in many ways for producing many high-quality and delicious products. Essentially, there are two major issues with milk production: first its high cost of production and second its quick perishability. These issues lead to an increase in the cost of transportation. Hence, high-tech operations are suggested for its production and transportation in order to keep its quality intact. As a result, milk producers are being linked to milk processing cooperatives domestically and internationally through liberalized international dairy markets, which is a great opportunity for milk producers of developing countries. In developed countries, the success of dairy industry is owed to cooperatives. Milk cooperatives provide a platform to many dairy farmers for their best interest, and it not only guarantees the purchase of their milk produced on a daily basis but also strengthens their bargaining power toward processing agents (FAO 2005).

Developing countries play an important role in the production of dairy products, and they are expected to contribute major chunk (around 80%) to global increase in the production of milk. Asia is a major contributor in this regard with 70% of total increase in milk production. India and Pakistan are main stakeholders in this regard. India has the largest growth rate in milk production with a 25% share in world production. It is followed by Pakistan having an annual growth rate of 2.5% and global share of 7% by 2027. They are jointly anticipated to provide one-fourth of global milk production by the end of 2027 (OECD/FAO 2018). In developing countries like India and

1 Historically "the White Revolution" term was first used in 1963 for Iran's White Revolution that was basically about redistribution of land from rich people to peasants. It is quite different from the discussion we are doing here.

Pakistan, there is a great potential for dairy farmers to compete regionally and internationally. Domestically, their governments have started initiatives like India's Operation Flood Program and Pakistan Dairy Development Company (PDDC) with the motive of bringing dairy development in elegant words "the White Revolution."

5.4.1 Indian White Revolution: Operation Flood

"Operation Flood" was an initiative of the Indian government that was a program designed for rural development started after the GR in 1970. The main focus of this program was the development of dairy production, which successfully brought "the White Revolution" in India by making it the largest milk producer. This revolution helped combat poverty in rural areas by discouraging malpractices by stakeholders in milk chain. It helped dairy producers to optimally utilize their resources for their betterment. A direct connection was developed between milk producers and consumers through the formation of the "National-Milk-Grid," which controlled price fluctuations and ensured producers a major share in the final price paid by the consumer. The Operation Flood empowered milk producers with modern technology and better management skills through the formation of milk cooperatives. The function of such milk cooperatives was to procure milk and provide services to bring a "Milk flood" by balancing prices and increasing rural remunerations besides increasing milk supply. This "Milk Flood" was ultimately named the White Revolution.

The Operation Flood was operationalized in three phases. Life span of Phase I was 10 years (1970–1980), where milk producers were linked with consumers in four major cities (Mumbai, Chennai, Delhi, and Kolkata); thus, mother dairies were established. Phase II was of about five years (1981–1985), where the number of milk sheds was increased from 18 to more than 130 (136 in total) and milk outlets were extended to almost 300 markets in cities. At the end of the second phase, 43,000 village cooperatives established 4.25 million milk producers. This self-sustaining system resulted in huge milk production as well as production of milk powder in thousands of tons by 1989. In Phase III (1985–1996), dairy cooperatives started expanding through infrastructural development and women empowerment. Currently, the dairy sector of India is performing very well due to ever-developing milk cooperatives. Verka, Parag, and Saras are a few examples of efficiently working milk cooperatives producing a wide range of milk value-added products at mega level besides providing milk and milk products.

5.4.1.1 Merits and Demerits of the Indian White Revolution

The White Revolution helped dairy farmers stratify their livelihood in a decent way by efficiently utilizing land resources. It increased their milk yield, and proper research and development helped improving health of animals and their nutrition. It transformed the dairy sector into a commercial activity and brought success to society. It was not mere mass production but mass production by masses. It made India from a milk importing country to the largest milk producer and one of the top milk exporting nations. Modernization of dairy industries and advanced infrastructures were available where millions of farmers earned enough revenue by doing dairy farming and rearing genetically improved milking animals.

Demerits of the White Revolution included noncooperative behavior of bureaucrats and cumbersome procedures in various states of India. The vested interest of higher-up's obstacles also made it difficult to get the whole benefit of the White Revolution. It resulted in the mismanagement of dairy units where the full benefits of efficient production could not be reaped. Working capital was lacking in some areas, and plants were not capable of bearing the operating expenses of dairy plants. Indian breeds became endangered due to overemphasis on foreign breed milk animals. Foreign breeds need specific temperature and feed to get adjusted to some new environment. It is also argued that Operation Flood replaced many desirable investments in research, development, and extension of agricultural enterprises other than dairy. Operation Flood also failed to address the problems of very poor farmers who were unable to afford the expenses associated with adopting innovation brought by it. Nevertheless, the White Revolution has brought major improvements in average living standards of rural Indian masses. Associated cooperatives played major role in bringing prosperity and development of rural economy of India.

5.4.2 Pakistani White Revolution: Doodh Darya[2] (The River of Milk)

Pakistan being heavily dependent on agriculture needs to feed its fast-growing population in an effective way. Livestock is an important subsector of Pakistan's agricultural economy with more than half contribution in agricultural GDP share. It contributes around 60% to agricultural value added. It also plays a significant role in national economy by contributing 11.5% in the national GDP. It is worth mentioning that this contribution has kept rising since the 1990s (GOP 2018). Around eight million households

2 PDDC (2006).

are involved in dairying and deriving more than one-third of their household income from it —an important source of employment and income in rural areas, aiding in reducing income inequality. Normally, small farmers have livestock for meeting their domestic need of daily milk/milk product consumption and cash generation. Nevertheless, having livestock at household is considered as an asset to be utilized in the case of any emergent cash requirement. It is essential to their livelihood, and it plays a significant role in reducing poverty. Hence, there are prospects for further development of this subsector as a significant part of overall economic development.

The bright role of the livestock sector is evident from developments in the dairy sector. The contribution of milk and its products is important not only for livestock sector but also for the agricultural economy as a whole. Pakistan stands fourth after China, India, and United States of America in the milk production (Rehman et al. 2017), whereas three-fourths of fresh milk are produced in Punjab province (Burki and Khan 2011; PDDC 2006). According to FAOSTAT (2022), more than three-fourths of milk production are contributed, 33% is provided by cows and approximately 3% is available from goats and sheep. Estimated annual gross milk production in 2017–2018 was 58 million tons out of which more than 80% (46.68 million tons) was available for human consumption. It has continuously increased over the period 2000–2001 to 2017–2018 with an average of around 36 million tons (Table 5.1).

Being part of food items, milk is the second important commodity after cereals from a consumption point of view. It is the top preferred commodity being consumed in Punjab (Naheed and Hussain 2014). Milk is supplied mainly through three types of milk markets, namely rural, urban, and international markets. Similarly, type of marketing chains can also be categorized into rural marketing, urban marketing, and international marketing chain. In rural marketing chain, milk is distributed in its raw form to the ultimate consumer through relatively smaller than other types of milk chains, mostly within the territory of that village (the chain is simply the direct farm shed–consumer link, or farmer-household in neighborhood). A more complicated chain might include multiple intermediaries like village retailer/milk man/village milk collector/rural consumer and skimming/contractor/bakery/retailer/processor/ultimate consumer. The latter form of intricate market chain is also found in urban and peri-urban areas, but the source seems to be rural milk producers. Difference between rural and urban value chain is accessibility to urban consumers and more control over the supply of milk in the case of urban marketing chain, which ultimately results in comparatively higher prices in urban value chain. Urban producers can rely on their own resources instead of involving middlemen in

Table 5.1 Milk production for human consumption in Pakistan (million tons).

Fiscal year	Milk for human consumption
2000–2001	26.284
2001–2002	27.031
2002–2003	27.811
2003–2004	28.624
2004–2005	29.438
2005–2006	31.970
2006–2007	32.996
2007–2008	34.064
2008–2009	35.160
2009–2010	36.299
2010–2011	37.475
2011–2012	38.617
2012–2013	39.855
2013–2014	41.133
2014–2015	42.454
2015–2016	43.818
2016–2017	45.227
2017–2018	46.682
Average	**35.829**

Source: GOP (2018).

selling milk to consumers. Similar is the case with peri-urban marketing chain. Processed marketing chain needs further development as in Pakistan only a meager proportion (3%) of milk is being processed due to the reason that dairy farms mostly smallholding lack modern facilities of cold chain (GoP 2022) and consumers' preference for loose milk as the price of processed milk is comparatively high. It is however encouraging that the government of Pakistan and international donors are taking a keen interest in developing milk processing industry. To modernize and develop dairy subsector, Dairy Processing and Infrastructure Development Fund (DIDF) has been developed in 2017–2018. The fund operated through the National Dairy Development Board (NDDB). Examples of public–private partnerships in this regard include the initiative of PDDC and the Livestock and Dairy Development Board (LDDB).

The government of Pakistan has initiated Livestock Loan Insurance Scheme (LLIS) in order to improve access to finance to Livestock and dairy sector and mitigate the risk of livestock loss because of disease, natural calamities, and accidents. LLIS was launched in 2013 and covers all sorts of livestock loans up to PKR 5 million for the purchase of animals. Under LLIS, the farmers are covered against animal death because of disease or natural death, flood, rainfall, storms, and accidental death (GOP 2018).

Commercialization in the dairy industry of Pakistan is believed to flourish given the high rate of urbanization, ever-changing preferences of consumers, and many other demand-driven factors affiliated with this value chain. It is evident that Pakistani consumers' demand has been increasing for milk, and its products and preferences of consumers are shifting toward processed and UHT milk. Factors responsible for such trends include high population growth rate, household income, nutritional needs, and awareness about the fact that such dairy products are less perishable than unprocessed ones (FAO 2011; PDDC 2006). Hence, dairy supply chain needs modification with the passage of time due to multiple pull and push dynamic factors. It would result in improved efficiency of milk and milk product's production and processing (Younas 2013).

The term "Doodh Darya" was introduced by PDDC in 2005 when it was formed. Currently, the White Revolution is taking place with the help of such governmental bodies and private sectors. Both public and private institutions are engaged in field through some projects including Milk Packaging Project in Central and Southern Districts of the former Northwest Frontier Province by NWFP's Livestock and Dairy Development Department, and Community Empowerment Through Livestock Development and Credit (CELDAC).

The project "Milk Packaging Project" adopted a bottom-up approach and developed the dairy industry of the former NWFP through partnership between public and private sectors. The project enhanced milk productivity through efficient milk marketing. The project was particularly focused on improving breeding, animal health, and feed offered to animals. It included training for women, farmers, and extension workers. It established milk collection and processing units and developed market links. The project formed farmer associations in villages, milk collection centers which were equipped with a cooling tank. Due to these efforts, income of farmers increased and rural poverty in target villages was ultimately reduced (Zia 2009). The UNDP's CELDAC project is another example in this regard which was completed in partnership with Nestlé and Engro, Pakistan. It was focused on women who play central role in livestock rearing and dairy-related activities. Women were technically trained in milk collection zones of Nestle and

Engro. Women were particularly benefited from this project. The project also provided credit for enterprise development.

Private sector organization like Nestle and Engro Foods of Pakistan are remodeling the rural economy of Pakistan through its dairy sector. They collect milk and milk products from rural farmers. They guide farmers to improve on-farm management of dairy to increase the yield and enhance the quality of milk. They guarantee timely payment for operations performed by farmers and ensure cost recovery better than they can manage to get from traditional methods (Sattar2018).

5.4.2.1 Challenges Associated with Pakistani Dairy Industry

Dairy farming in Pakistan differs from that in developed countries like Europe and the United States of America, where it is a separate enterprise. It is done alongside cropping just as a "by-product." This parallel application of two kinds of farming is described as mixed farming. In this way, farmers successfully benefit from cropping by utilizing crop residues as feed for livestock and animal waste as fertilizer. Integrated farming if done wisely may enhance the efficiency of farms. However, there are chances of reduction in the efficiency of dairy operations if taken in isolation where management techniques utilized solely for the production of milk and its value-added products are followed.

Milk is mostly being produced and managed at smallholder farmer's level in rural areas of Pakistan. Commercialization of milk at macro level is a challenge for Pakistan's rural economy. Bulk of milk supply comes from around 55 million smallholder farmers. And that by itself is less efficient than large scale because of economies of scale due to role of fixed capital costs. More than 80% of milk collected in the country comes from rural producers. The remaining 20% of milk is collected from peri-urban and urban producers (Zia 2009). Milk marketing is done mostly through informal agents, and only 3–5% of the country's total milk production is marketed through formal channels. The remaining 97% is produced and marketed in raw form by informal agents in the marketing chain (FAO 2013; GOP 2022).

Milk is mainly being produced by subsistence farmers who lack in proper management and animal breeding skills. This results in lower potential farm productivity and returns at national level. The number of animals in Pakistan is higher than in many developed countries; however, per-animal milk productivity is very low. According to the recent Pakistan Livestock Census (2006), the average milk yield of cows is 6.14 l/animal/day and that of buffaloes is 7.93 l/animal/day (GOP 2006). Milk production per animal is subject to availability of green fodder, which is present only during specific seasons. Dairy farmers do not only work on a small scale, but they are also less organized and mostly work in isolation instead of having cooperative setup, which results in low farm productivity and profitability in the country.

Another constraint for small daily farmers in Pakistan is the lack of proper access to financial assets like insurance and credit. Dairy farmers relay on informal source of credit, where written contracts are mostly missing. Their preference of relying on informal source of credit is for reason that it is hassle-free and requires no/less mortgage. A huge majority (80%) of small producers having less than five animals remain non-bankable as they do not fulfill the condition of holding more than five acres of land (SBP 2012). They are vulnerable to crises, especially in the case of death of animal and/or its injury. Lack of easy access to credit also prevents farmers from improving their business and access to markets. They normally must rely on middlemen to sell their produce. Middlemen exploit them by offering low prices, signing binding contracts of sales, and restricting the benefits of higher seasonal prices to themselves.

Milk loss due to a shortage of cold milk chain is another obstacle in dairy management. An estimated 15–20% of total milk production is lost due to unavailability of proper storage and processing facilities, and almost six million tons of milk is lost due to poor management skills of farmers (Farooq 2016). A dairy farmer operating at small scale is unable to afford a cold storage tank which is very expensive. If they manage to buy it, energy crises in rural areas would be an obstacle in operating an electric tank. If these small-scale dairy farmers are interlinked through cooperatives, cold chain infrastructure is an affordable way of improving milk supply. Besides saving the cost of milk procurement to be paid by the milk processing firms, it also provides farmers with a means of accessing processors directly.

The lack of price regulations for both input and output in the case of milk is another obstacle in dairy development. Provincial Food Departments should be allowed under law to declare milk as food commodity and empower a District Price Review Committee which regularly reviews milk prices in respective districts. The committee should consist of stakeholders from the livestock department, farmers, dairy retailers, and consumers (Zia 2009).

Before replicating international models in Pakistan, consequences of innovations should be kept in mind. For instance, Pakistan's few medium to large-scale farmers on persuasion of an international corporation imported high-yield cattle from Australia. That breed could not tolerate heat stress in Pakistan and died, which resulted in huge loss for farmers, since there was no insurance on animals (FAO 2011). This is a challenge for the government to look into such issues while keeping the socioeconomic and financial consequences associated with such initiatives.

5.4.2.2 Way Forward for "Doodh Darya"

To address the aforementioned web of interrelated challenges, several strategies are recommended to be followed for dairy development as discussed in the following section.

In developing countries, establishment and financing of milk cooperatives are highly required. In Pakistan, where almost one-fourth of milk is wasted due to faulty transportation and lack of cold chain (GoP 2022). In Pakistan's case, there exists a gap between the profitability of small-scale and large-scale dairy farmers. Smallholder farmers cannot avail economies of scale and market benefits. They may enjoy such benefits if grouped into corporates and provided with different incentives like tax evasions/reductions subsidies on inputs, access to credit especially microfinance, and required market infrastructure. Corporates enjoy capital accumulation, mechanization, and enhanced fodder storage facilities. However, corporate farming demands high investment in infrastructure and capital. Thus, corporate farms require advanced technology and high-quality fodder for preparing livestock feed. For maintaining and enhancing the quality of milk in bulk, cold storage facilities are a prerequisite. Again, corporates may access such facilities because of having good access to credit. Government needs to encourage the local manufacturing industry for producing low-cost cold storage equipment for dairy farmers. Corporates have better access to subsidized electricity and market orientation when compared with small farmers. Affordability of fodder for raising livestock is a challenge for small-scale dairy producers. Share of fodder cost in total cost of dairy farms is around 38% (Burki and Khan 2016). The infrastructure required for silage preparation at farms is mostly unaffordable for many smallholders' dairy producers. Efforts to decrease the cost of production in dairy sector require taking this fact into consideration. Microcredit may be provided to smallholders to aid them for installing such infrastructure for storing fodder during peak seasons for silage making.

The corporate dairy farming is in its initial phase. Fields needing improvement by them include reducing productivity cost (for instance, by hiring local experts instead of foreigners). They also need to focus on utilizing different methods of artificial insemination and selective breeding, installing better animal housing facilities, and lower fodder costs by producing animal feed at farms. Strategies like tagging animals to grade them according to milk yields are required to be part of farming practices. Farm management practices play a role in the quality of milk. Educational programs play an important role particularly in rural areas, for training small farmers regarding improved dairy farm management. Veterinary professionals may be hired to guide on improving milk quality and hygiene of animals.

Government needs to play its role for encouraging research on new dairy technological practices and sustained engagement between dairy chain partners (Yaseen 2015) and enhance awareness regarding value-added milk products. Studies on prevalent consumption patterns may help the milk processing industry shape their marketing strategies. Additionally, programs on education and training of farmers are required to be encouraged so that farmers are trained enough regarding milk quality specification, and they have provided them with necessary milk testing equipment. Vocational training and extension services administered to smallholder dairy producers on animals' feed requirements, high-quality indigenous sources of feed, and optimal mixtures of ingredients can go a long way toward improving basic skill set of the target groups. Extension workers if well trained and provided with sufficient incentives may train farmers and educate them regarding improved management and production of dairy products and accessing financial resources. Improved and sustainable breed programs implemented by various corporates already working in the country had great experience by having enhanced milk productivity. There is a dire need to increase their bargaining power and encourage them to stay in dairy sector and produce milk in bulk to be collected by corporates.

The corporates have connections with milk processing industry. There exist various issues along the supply chain (farmers' milk processing industry). Milk processing industry creates demand for dairy farmers and corporates. Nonetheless, milk processing industry is not operating to its fuller potential. For the progress of milk processing industry on sustainable basis, then an integrated approach is required to be adopted for solving issues of both suppliers of raw milk (dairy farmers) and the processing sector itself. One way of financing such inputs to milk farmers is through intermediaries, following India's model, where cooperatives and intermediaries played a key role in bringing the White Revolution. To start a corporate setup in dairy sector, massive investment in infrastructure and instruments is required. It is recommended that the central bank plays its role in financing these inputs using different financial tools. Utilizing the tool of microfinance is imperative in this context for providing the required capital and enhancing the total factor productivity of small-scale dairy farmers, which would also aid in solving issues related to milk processing industry.

To encourage profitability in dairy farming, laws related to price control should be reviewed and implemented. A level playing field should be set while doing so. Similarly, planning and decision-making should be based on improved access to information where annual national livestock census should be conducted by the government of Pakistan. Detailed analysis should be conducted for improved decision-making at macro-level. Dairy information should be accessible through a GOP-run data repository. In this way, the government would be able to guide the use of modern information technology and improve access to market information.

The government of Pakistan should invest in research and development to promote good packaging of dairy products and their development and marketing. The dairy industry should be modernized and become competitive in the world. For instance, PDDC can be used as a platform for gathering smallholders and linking them to different market channels, especially the formal ones. Government and other nongovernmental actors should help this setup by providing infrastructure and other facilities and enable milk producers to meet international standards of milk quality and milk supply timetables as well as to label their products (milk and milk products). Furthermore, the government should be proactive in providing institutional arrangements for the functioning of dairy markets. Policymakers of Pakistan should follow some practical action plan for the improvement of dairy industry. It should support dairy producers by providing facilities like shared dairy freezers, milk procurement centers, and cooling units. Farmers should be facilitated to visit modern dairy industries for getting acknowledged to modern techniques. Dairy networks should be created where producers may easily approach processors and consumers.

5.4.3 Brazilian Experience of Dairying

Brazil is the third largest producer of milk. It has the world's second largest dairy herd. It has remained primarily a net importer of milk products, and it transformed to net exporter since it liberalized its trade back in 1993. Growth of milk production has been witnessed since 1993 from 15.6 billion liters to 36.8 billion liters in 2020 (FAO 2022). Brazil is that part of Latin America where livestock is being reared with the purpose of milk production in its southwest and southern parts (for instance, Rio Grande do Sul and Santa Catarina), where weather conditions are quite favorable and inputs required for milk production are low cost (Spers et al. 2013). In the northern parts of Brazil, there is rise in production of meat causing a threat to the Amazon Forest Scale of distribution of milk produced in Brazil is diverse with respect to size of farm, type of dairy farmer, type of animal, and production techniques. Milk producers belonging to different profiles have access to dairy knowledge, financial facilities, consumers, and other actors of the chain. In rural areas, dairy is a family business both for subsistence and income earning. Dairy industries are large and diverse in nature with private sector playing a key role, whereas producers are relatively weaker

(FAO 2013). A number of dairy producers in Brazil are reportedly around 1.2 million, ranging from small farmers with yield per animal less than 10 l/day milk production not by using any advanced technology to large dairy producers having sales of more than 60,000 l/day. If compared with India, where number of dairy farms is much higher but average milk yield is around 12 l/animal/day of foreign breeds, and that of local breeds is even lower (GOI 2021) which is even lower in the case of Pakistan. Nonetheless, in the case of Brazil, smallholders (<50 l/day) are only responsible for 20% of national milk production. More than 60% of total milk is being supplied by large dairy farmers with the production capacity of 200 l/day, but they constitute only 3% of total daily producers (Douphrate et al. 2013). Brazil is self-sufficient in the production of milk and milk products, but it is striving to increase herd size slowly to meet domestic demand for milk and milk products by being in pace with population and income of its inhabitants. Success of Brazilian dairy industry can be attributed to sensible planning and support from government. Farmers participate in dairy development programs by government for increasing productivity through the improvement of pasture and animal's improved genetic basis.

A huge population of this country is employed in its dairy value chain. Cooperatives have a central role to play in the value chain of dairy in the country. The example of Brazil, however, is different from India and Pakistan. Cooperatives were developed based on vertical integration for promoting industrialization of dairy products. The vertical integration strategy of cooperatives was predominantly followed in the 1980s. In the modern era, this setup changed, and new cooperative organizations were developed known as "Virtual Cooperatives." These virtual organizations make Brazil's case different from that of India and Pakistan. These virtual cooperatives are mostly recognized by their small number of associated rural farmers, who do not possess a building or who are not capable of bearing the management cost of traditional cooperatives. These cooperatives ensure quality standards and networking among producers. These cooperatives neither have headquarters nor industrial parks.

Success of dairy in Brazil is predominantly because of technological innovations in this sector. The use of technology not only has increased production of quality and cleanliness of milk and milk products but also increased consumers' trust on dairy products produced thereby. For instance, in Brazil, robots and/or machines are used for milking. This allows the milking of several cows simultaneously. This process ensures cleanliness and sanitization of the surrounding environment, and the quality of milk extracted this way is better than manual extraction, where chances of contamination are high. Dairy farms use sensors, which allow collection and sharing of data with each other. These sensors can measure nutritional contents and presence of antibiotics in milk. Using such digital transformation of dairy sector in Brazil helped people link to this value chain allowing updating and sharing of information about animals at dairy farms. It helps them in many ways to take necessary actions for better productivity and quality. Capturing dairy data allows farmers to directly access data. It also helps other farmers who use these technologies wisely in their production.

5.5 Conclusions

For developing countries, agriculture plays an imperative role in ensuring food security and sustaining the livelihood of the masses. History reveals that there have occurred several revolutionary changes in agriculture. The most prominent feature of all transformations over time was the dynamics of agriculture from being a source of subsistence to a commercial activity. The first ever agricultural evolution was the Neolithic Revolution, characterized by the transition from hunting food to domestication. Neolithic humans domesticated animals predominantly for getting milk and meat, which added some variety to their diet. Nevertheless, the majority of masses had dietary homogeneity given their substantial reliance on a limited range of food items. The Second Agricultural Revolution was "the British Agricultural Revolution" of mid-seventeenth century marked by extraordinary increase in agricultural production because of improved total factor productivity, technological advancement, and crop rotation. This led to the Industrial Revolution, where mega food chains and corporations were formed resulting in decreased prices of food and an overall increase in food accessibility and utilization. The Industrial Revolution remained a success; however, a staggering proportion of global population was still hungry. This represents the famous Malthusian dynamics, where food per person does not increase over time because of an exponential increase in population that outpaces food production. The third agricultural revolution was the Green Revolution of 1960s. It was initiated with the motivation of bringing food security and reducing rural poverty. It was somehow a more planned set of agricultural technology transfer initiatives, plant breeding, and genetic engineering that enhanced agricultural productivity. The GR led to increased area under crops and high productivity of crops, particularly cereal crops. As a result, per capita food consumption increased, but there was still a need to improve the nutritional status of their population. The only aim of the green revolution was to increase food items' production and make it sufficient to feed everyone. It did not consider issues like nutritional insecurity, environmental concerns, and income and gender inequalities. On the positive, the GR improved average conditions, but on the negative, it increased various types of inequalities. The reason was that the key beneficiaries were progressive farms instead of the poor ones (Andersen et al. 1985; Pingali 2012a). Thus, benefits from the GR were not evenly distributed, which increased the poverty gap and inequality between rural and urban areas of the developing countries. The post-Green Revolution has been characterized by substantial changes in food production and consumption patterns. Nutritional security became more important than mere food security. Out of several failures of the GR, micronutrient malnutrition was a significant one (Pingali 2012b), where decreased price of staple food items discouraged the productivity of micronutrient-rich crops like vegetables and pulses (Gomez et al. 2017).

The White Revolution is the fourth agricultural revolution representing success of dairy farming. Developing countries, particularly Asia, contribute to most of the global milk production. The Indian White Revolution is exemplary success that has transformed India from milk importing country to the largest milk producer and one of the top milk exporting nations. Besides India, Brazil is also prominent due to its dairy farming. It is the third largest producer of milk. It has world's second largest dairy herd. Brazil's "Virtual Cooperatives" make Brazil's case different from other milk producing countries. These virtual cooperatives are mostly recognized by their small number of associated rural farmers who do not possess a building or who are not capable of bearing management cost of traditional cooperatives. These cooperatives neither have headquarters nor industrial parks. Additionally, dairy farms utilize technological innovations, for instance, robots and/or machines are used for milking. Sensors are used for collection and sharing of data with each other. These sensors can measure nutritional contents and presence of antibiotics in milk. Using such digital transformation of dairy sector in Brazil helped people linked to this value chain allowed to bring updating and sharing of information about animals at dairy farms. Pakistan though one of the top producers of milk, it processes only a meager proportion (3%) of milk as dairy farms mostly being smallholding lack modern facilities of cold chain (GoP 2022) and consumers' preference for loose milk as price of processed milk is comparatively high. Almost one-fourth of milk is wasted due to faulty transportation and lack of cold chain (GoP 2022). There exists a gap between profitability of small-scale and large-scale dairy farmers. Corporate dairy farming is in its initial phase. If these small-scale dairy farmers are interlinked through cooperatives at large scale, cold chain infrastructure is affordable way of improving milk supply. Corporates have connections with milk processing industry. Nonetheless, milk processing industry is not operating to its fuller potential. For the progress of milk processing industry on a sustainable basis, an integrated approach is required to be adopted for solving issues of both suppliers of raw milk (dairy farmers) and the processing sector itself. To start a corporate setup in dairy sector, massive investment on infrastructure and instruments is required. It is recommended that the central bank plays its role in financing these inputs using different financial tools like microfinance. Integrated farming, if done wisely may enhance the efficiency of farms. Government needs to encourage local manufacturing industry for producing low-cost cold storage equipment for dairy farmers. Research on new dairy technological practices and sustained engagement between dairy chain partners is highly required. Pakistan needs to learn from India's model where cooperatives and intermediaries played a key role and the model of Brazil where "Virtual Cooperatives," "robots," and technology have brought "the White Revolution."

References

Andersen, P. and Peter, B.R.H. (1985). The impact of the green revolution and prospects for the future. *Food Rev. Int.* 11: 1–25.

Burki, A.A. and Khan, M.A. (2011). Formal participation in a milk supply chain and technical inefficiency of smallholder dairy farms in Pakistan. *The Pakistan Development Review* 50 (1): 63–81.

Burki, A.A. and Khan, M.A. (2016). *Pakistan's Dairy Sector: Lessons from the Past to Build a Resilient Dairy Industry. A Technical Report*. Pakistan: Lahore University of Management Sciences (LUMS).

Chand, R. and Raju, S.S. (2009). Instability in Indian agriculture during different phases of technology and policy. *Ind. J. Agri. Econ.* 64 (2): 187–207.

Dorjee, K., Broca, S., and Pingali, P. (2003). *Diversification in South Asian Agriculture: Trends and Constraints*. Rome, Italy: Food and Agriculture Organization of the United Nations.

Douphrate, D.I., Hagevoort, G.R., Nonnenmann, M.W. et al. (2013). The dairy industry: a brief description of production practices, trends, and farm characteristics around the world. *Journal of Agromedicine* 18 (3): 187–197.

Evenson, R. and P. Pingali (2007). Overview. Handbook of Agricultural Economics. 3 (eds R. E. Evenson&P. Pingali). Amsterdam, The Netherlands: Elsevier Science.

FAO (2005). Developing countries and the global dairy sector part/ global overview. Pro-Poor Livestock Policy Initiative. Working Paper 30 (by Knips, V.). Rome, Italy.

FAO (2011). World Livestock 2011—Livestock in Food Security. FAO, Rome.

FAO (2013). Milk and Dairy Products in Human Nutrition. Muehlhoff, E., A. Bennett and D. McMahon. Food and Agriculture Organization of the United Nations, Rome, Italy.

FAO (2019). *Food and Agriculture Organization of the United Nations Stat (FAOSTAT)*. Rome, Italy: Food and Agriculture Organization of the United Nations.

FAOSTAT (2022). *Food and Agriculture Organization of the United Nations Stat (FAOSTAT)*. Rome, Italy: Food and Agriculture Organization of the United Nations.

Farmer, B.H. (1981). The "Green Revolution" in South Asia. *Geography* 66 (3): 202–207.

Farooq, U. (2016). Milk supplies in Pakistan: Issues and challenges facing the dairy economy. In: *A paper presented in Conference on 'Milk Supplies in Pakistan: Quantity, Quality and Legal Framework*. Islamabad: COMSATS Institute of Information Technology (CIIT).

Feder, G., Just, R.E., and Zilberman, D. (1985). Adoption of agricultural innovations in developing countries: a survey. *Econ. Dev. Cultural Change.* 33 (2): 255–298.

Fischer K.S., V.G. Cordova (1998). Impact of IRRI on rice science and production. In: Pingali P.L., Hossain M., eds. The Impact of Rice Research. Manila, Philippines: International Rice Research Institute and Thailand.

Fitzgerald-Moore, P. and Parai, B.J. (1996). The green revolution. Research paper in progress. http://www.ucalgary.ca/~pfitzger/green.pdf. (accessed 08 November 2023).

Flavin, C. and Young, J.E. (1993). Shaping the next industrial revolution. In: *State of the World, 1993* (ed. L.R. Brown et al.), 180–257. New York: W. W. Norton & Co.

Ganzel Group (2007). Farming in the 1950s and 60s: The Mexican Agricultural Program. Wessels Living History Farm. http://www.livinghistoryfarm.org / farminginthe50s/crops_14.html (accessed 28 October 2022).

Ghassemi, F., Jakeman, A.J., and Nix, H.A. (1995). Pakistan. In: *Salinisation of Land and Water Resources: Human Causes, Extent, Management and Case Studies*, 369–395. Sydney: University of New South Wales Press Ltd.

GOI (2021). Department of Animal Husbandry and Dairying – Annual Report 2020–2021.

Gollin, D., Morris, M., and Byerlee, D. (2005). Technology adoption in intensive Post-Green Revolution Systems. *American Journal of Agricultural Economics.* 87 (5): 1310–1316.

Gomez, D.J., Twyman, R.M., Zhu, C. et al. (2017). Biofortification of crops with nutrients: factors affecting utilization and storage. *Curr. Opin. Biotechnol.* 44: 115–123.

GOP (2006). *Pakistan Livestock Census 2006*. Islamabad: Agriculture Census Organization, Statistics Division.

GOP (2018). *Economic Survey of Pakistan (Various Issues)*. Islamabad, Pakistan: Finance division. Government of Pakistan.

GOP (2022). *Economic Survey of Pakistan (Various Issues)*. Islamabad, Pakistan: Finance Division. Government of Pakistan.

Hazell, P. (2009). *The Asian Green Revolution*. Washington DC, United States of America: International Food Policy Research Institute.

Hicks, N.L. (2011). *The Challenge of Economic Development: A Survey of Issues and Constraints Facing Developing Countries*. United States of America: AuthorHouse.

Joshi, P.K., Gulati, A., and Cummings, R. (2007). Agriculture diversification in South Asia: beyond food security. In: *Agriculture Diversification and Smallholders in South Asia* (ed. P.K. Joshi, A. Gulati, and R. Cummings Jr.), 48. New Delhi.: Academic Foundation.

Khan, M.A. and Shah, A.A. (2009). Sustainable development of agriculture in South and Central Asian Regions – a key to prosperity. *Central Asia Journal*.

Kiple, K.F. and Ornelas, K.C. (2000). *The Cambridge World History of Food*. Cambridge University Press.

Kohler, S. (2007). The Green Revolution, The Rockefeller Foundation, 1943. In: *In Casebook for The Foundation: A Great American Secret*, 51–58. United States of America: Public Affairs.

McIntyre, B.D., Herren, H.R., Wakhungu, J., and Watson, R.T. (2009). *International Assessment of Agricultural Knowledge, Science and Technology for Development Synthesis Report: A Synthesis of the Global and Sub-global IAASTD Reports*. Island Press.

Naheed, K. and Hussain, I. (2014). Food consumption pattern in Pakistan: recent interprovincial empirical evidence. *Gomal University Journal of Research*. 30 (1): 38–43.

Nelson, A.R.L.E., Ravichandran, K., and Antony, U. (2019). The impact of the Green Revolution on indigenous crops of India. *J. Ethnic Foods* 6 (8).

OECD/FAO (2018). OECD-FAO Agricultural Outlook 2018–2027. Organisation for Economic Cooperation and Development (OECD) and the Food and Agriculture Organization of the United Nations (FAO).

PDDC (2006). *The White Revolution: "Doodh Darya"*. Lahore: Pakistan Dairy Development Company.

Pingali, P. L. (2012a) The Green Revolution and crop diversity (Chapter 12). In D. Hunter, L. Guarino, C. Spillane, and P. McKeown (Eds.) Handbook of Agricultural Biodiversity. Routledge, New York.

Pingali, P.L. (2012b). Green revolution: impacts, limits, and the path ahead. *Proc. Natl. Acad. Sci. U. S. A.* 109: 12302–12308.

Postel, S. (1989). *Water for Agriculture: Facing the Limits*. Washington (DC): Worldwatch Institute.

Rehman, A., Jingdong, L., Chandio, A.A., and Hussain, I. (2017). Livestock production and population census in Pakistan: determining their relationship with agricultural GDP using econometric analysis. *Inform. Process. Agric.* 4: 168–177.

Sattar, E. (2018). Dairy Farming in Pakistan: The Transformation of Small Farms and Livestock Management Practices. https://ssrn.com/abstract=3171075 (accessed 28 October 2022).

Scherr, S. and Satya, Y. (1996). *Land Degradation in the Developing World: Implications for Food, Agriculture, and the Environment to 2020*. Washington, DC: International Food Policy Research Institute.

Setboonsarng, S. (2006). Organic agriculture and the MDGs: impacts reported in existing literature. Discussion Paper No. 54. Tokyo: ADBI.

Shah, Z.-u.-H. and Arshad, M. (2006). *Land Degradation in Pakistan: A Serious Threat to Environments and Economic Stability*. Eco Services International.

Shiva, V. (1991). *The Violence of the Green Revolution*. London: Zed Books.

Singh, R. (2000). Environmental consequences of agricultural development: a case study from the Green Revolution state of Haryana. *India. Agric. Ecosyst. Environ.* 82: 97–103.

Smith, J.C., Ghosh, A., and Hijmans, R.J. (2019). Agricultural intensification was associated with crop diversification in India (1947–2014). *PLoS ONE* 14 (12).

Spers, R.G., Wright, J.T.C., and Amedomar, A.D.A. (2013). Scenarios for the milk production chain in Brazil in 2020. *Relations Between Agriculture and Society* 48 (2).

State Bank of Pakistan (2012). *Report on Dairy Value Chain in Pakistan*. Karachi. Pakistan: Agricultural and Microfinance Department, SBP.

Yaseen, A. (2015). Collaborative innovation in the Pakistan's dairy industry: effectiveness of managerial leadership and organizational readiness. PhD Thesis.

Younas, M. (2013). *The Dairy Value Chain: APromoter of Development and Employment in Pakistan*. Kassel, Germany: International Center for Development and Decent Work.

Zia, U.E. (2009). Pakistan: a dairy sector at a crossroads. In: *Smallholder Dairy Development Lessons Learned in Asia. RAP Publication 2009/2*, 76–92. Bangkok, Thailand: APHCA.

6

Dynamics of Income and Income Redistribution in Developing Countries

Rakhshanda Kousar[1], *Muhammad Sohail Amjad Makhdum*[2], *and Khalid Mushtaq*[1]

[1] *Faculty of Social Sciences, Institute of Agricultural and Resource Economics, University of Agriculture, Faisalabad, Pakistan*
[2] *Department of Economics, Government College University, Faisalabad, Pakistan*

6.1 Historical Background

Income distribution and income inequality are crucial issues at the global level. It is rising in almost 50% of the world, including less developed, low- and middle-income countries. Several countries in Asia and Africa have seen tremendous rises in income inequality. The countries with the highest rate of inequality have observed a disproportionate effect on food and nutrition security in a time of economic crisis because they spend a major portion of their income on food items. Income inequality tends to enhance the likelihood of severe food insecurity, and its effects are 20% more in low-income countries than in middle-income countries (WHO 2019).

Until the end of World War II, development economists had focused on growth and poverty in developing countries, as reduction of poverty had been a universal priority for more than four decades; the long-term impact of the importance of poverty is revealed by the adoption of the Sustainable Development Goals and the UN commitment to target poverty reduction (Nino 2015). They also focused on savings, investment, and foreign aid, but they did not give due importance to income distribution. The decline in the vulnerability and length of poverty greatly depends on the ability to target poor people in society (Krantz 2001) and the causes of poverty (Krishna 2007). The availability of panel data sets has permitted analysis of the dynamic and

Food Security in the Developing World, First Edition. Edited by Muhammad Khalid Bashir, Steven G.M. Schilizzi, and Ghaffar Ali.

important nature of poverty at the household level in the past decades (Baulch 2011; Dartanto and Nurkholis 2013; Spielman et al. 2017). Kuznets (1955) explained the deterioration in income distribution as a natural phenomenon in the development process when people shifted from subsistence agriculture to the modern industrial sector.

In the 1990s, development economists shifted their core concern toward political economy analysis due to the availability of facts and figures on income distribution in developing countries. Later, they became more interested in analyzing the effects of income distribution on economic growth. Easterlin (1995) reported that happiness in terms of well-being had a direct relationship with one's income and an inverse relationship with the income of others. Rajan (2010) highlighted the role of inequality in disturbing the financial cycle, which he regarded as the beginning of a crisis in an economy.

The consequences of inequality on growth and development are mixed in both the empirical and theoretical literature and have different effects on developing and developed economies. Moreover, little attention has been given by policymakers to addressing the issue of inequality in the context of growth and development.

Higher inequality in developing countries hinders the process of economic growth by reducing labor productivity, which hampers the buildup of human capital, a key ingredient for economic growth. For instance, Assa (2012) found a negative relationship between inequality and economic growth in developing economies, but the relationship was insignificant in the case of developed economies. Ostry et al. (2014) also found a negative association between inequality and economic growth, independent of the level of economic development. Carvalho and Rezai (2015) discussed both empirically and theoretically that increased income concentration could reduce aggregate demand and undermine growth as the richer tend to spend a lower fraction of their income than middle- and lower-income groups.

Although inequality directly affects economic growth, it can also lead to public policies that hinder growth and disproportionately benefit the poor (Bourguignon and Dessus 2009). Babu et al. (2016) showed a significant detrimental effect of inequality on long-term economic growth. Still, they could not find a trade-off between economic growth and redistribution in the short and long run.

The literature suggests that some degree of inequality benefits developing countries. It may stimulate poor people to excel, compete, save, and invest in improving their quality of life (Barro 2000). Berg and Ostry (2011) also showed many countries a positive linkage between equality and sustainable growth.

6.2 Concept and Measures of Income Distribution

Income distribution refers to the dispersion pattern of a nation's Gross National Product (GNP) among the different members of society within an economy. Income and income distribution are always a key concern for policymakers and remain an essential part of economic theory. Income distribution has been a focal point for political economics since the time of Adam Smith.

In the empirical and theoretical literature, economists focused on two prime measures of income distribution: the size or personal distribution and functional or factor share distribution in society. Classical economists like Adam Smith, Karl Marx, David Ricardo, and Thomas Malthus focused on the functional distribution of income between factors of production. This is mainly concerned with the allocation of income between the major factors of production (land, labor, capital, and organization).

However, at the end of the nineteenth century, attention was diverted toward the size distribution of income which draws a clear-cut difference between various economic classes, made possible by the availability of official and household data on income tax. The income distribution is considered a direct measure of social welfare that shows the amount of income received by rich, poor, and middle-class individuals or families. The shape of the size distribution of income represents the role and patterns of ownership that each factor has in the production process. It is mainly concerned with the total income of individuals and households but totally ignores the sources and contribution of income among each member. There exist some common measures of the size distribution of income. A few are the Pareto index, the standard deviation of logarithms of incomes, the coefficient of variation, the Lorenz curve, and the Gini concentration ratio (GCR).

6.2.1 Lorenz Curve

The most common graphical measure of size distribution is the Lorenz curve. The Lorenz curve represents the increasing proportion of income units relative to the increasing proportion of income obtained when these units are organized in ascending order of their income. The Lorenz curve has been primarily used as a suitable graphical tool to represent the size distribution of wealth and income. Recently, Atkinson, who presented a theorem relating to the Lorenz curve and the social welfare function, has renewed interest in the Lorenz curve. He explained that the position of income distribution according to the Lorenz curve criterion is equal to the ranking entailed by an aggregate economic welfare function of individuals, provided the Lorenz curves do not intersect (Atkinson 2015). The generalized Lorenz curves are described as concentration curves, while the Lorenz curve is just a particular case of such curves, viz., the concentration curve for income (see the Chapter 3, Measures of development by Singh, 2009)

To draw a Lorenz curve, the number of income recipients in cumulative percentages is arrayed along the horizontal axis. The vertical axis represents the percentage of income received in cumulative population percentages. The figure is a square shape, and the diagonal is drawn from the lower left corner (0% of income received by 0% of recipients) to the upper right corner (100% of income received by 100% of recipients). If every person receives the same income level, then the Lorenz curve would lie along the 45° lines, which shows perfect equality. The farther the Lorenz curve is from the 45° lines, the greater the income distribution inequality. If one person receives all the income, the Lorenz curve will look like a backward L, just like the horizontal line would fall at the bottom with 0% of income, and the vertical line is up on the right-hand side. The vertical line will show that 100% of families receive 100% of their income. Although it can compare income distribution over time and across space, it fails to quantify the inequality (Figure 6.1; Singh 2009).

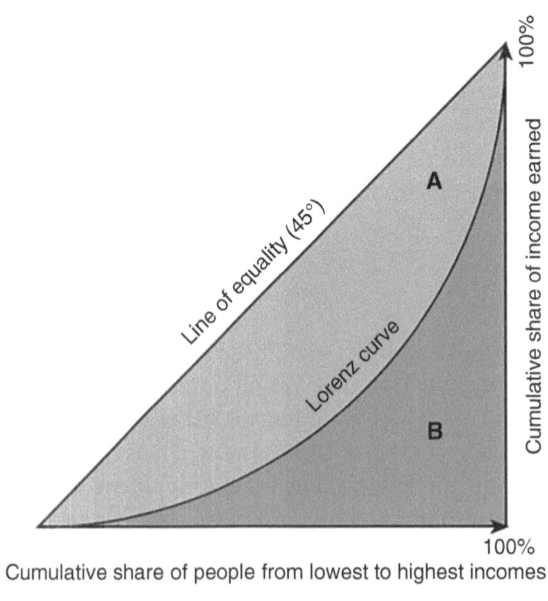

Figure 6.1 Income distribution represented by the Lorenz curve.

6.2.2 Gini Index

It is the most used measure of inequality. It measures the extent to which the distribution in the economy deviates from the perfectly equal distribution. The Gini index is the ratio of the area between the Lorenz curve and the 45-degree line to the area under the 45-degree line. A high value of the index shows an unequal distribution. The advantage of the index is that it makes an easy comparison of the income distribution of the two populations irrespective of their sizes. But the index is not free of limitations. The Lorenz curve shows the difference in income inequality, but it cannot quantify it.

Gini coefficient level varies considerably in many countries and shows significant changes over time. There is no trend in inequality levels over time as some countries show an increase in the Gini coefficient and some countries show a falling trend between the two points in time considered. Comparing the Gini index between two time periods does not indicate a long-term trend, but the magnitude of change can say something about inequality (Mckay 2002). The Gini coefficient can be calculated as:

$$\text{Gini coefficient} = \text{Area}\left(A\right) / \text{Area}\left(A + B\right)$$

6.2.3 Gini Concentration Ratio

The GCR is derived from the Lorenz curve. It estimates how the income distribution among members of an economy diverges from a perfectly equal distribution. It represents the proportion of the area lying between the diagonal and the Lorenz curve relative to the total area under the diagonal. The theoretical range of the GCR lies between 0 (perfect equality) and 1 (perfect inequality). It can also be calculated from the mean difference instead of the ratio. The formula for calculating the Gini coefficient ratio is

$$\text{GCR} = \sum_{i=1}^{n}\sum_{j=1}^{n}\frac{f_i f_j / x_i - x_j}{2\bar{X}N}$$

where n is the number of income classes, f_i and f_j are the frequencies in the ith and jth classes, x_i and x_j represent the mean incomes of ith and jth classes, N is the population, and $\bar{X}$ is the overall mean income.

Gini's concentration ratio G is defined in the cumulative distribution function. The formula is very easy to use and makes it easier to calculate values and examine the ratio's value. The GCR results from the Lorenz curve and is the index of inequality in a distribution. Kendall and Stuart (1979) provided the ratio and the general explanation of this curve. Modern economists and policymakers are more concerned with the income distribution pattern at the individual and household levels than before. The following section will explain the trends of income inequality across time and space to reveal a deeper understanding of income distribution.

6.3 Global Trends of Income Distribution

The key challenge of our time is widening the wealth gap between the wealthier and the poorer people of the same society. In recent decades, this income gap has been at its highest in developed economies, but this trend is mixed in developing and emerging economies (Obst 2015). To appreciate the extent and depth of poverty in the Third World, it is also essential to understand the increasing income gap between rich and poor individuals in less developed countries. Some emerging economies like China and India have exhibited declining income inequality trends in the last few decades as poverty has been reduced. These economies still face inequalities regarding access to necessities like health, education, and finance (Clements et al. 2015). Data on inequality shows different trends in less developed, emerging, and developed economies. For instance, as a measure of the degree of inequality, if we compare the share of national income to the bottom 40% of a country's population with the richest 20%, we find that countries like Brazil, the United States, Mexico, Ecuador, Kenya, Peru, Pakistan, the Philippines, and Malaysia have a substantial degree of income inequality. In contrast, emerging economies like China, India, and Chile, have moderate inequality, and Japan and Taiwan have relatively less inequality in income distribution. Figure 6.2 shows the inequality trends of some industrialized and developing economies.

Moreover, there is no obvious correlation between per capita income and inequality in income distribution. For example, the Philippines has almost the same average per capita income level as Taiwan but wider income inequality between the top 20% and the bottom 40% of its total population. This shows that development can be measured in terms of growth of total or per capita income but is represented in terms of welfare that individuals gain. Many exogenous factors (globalization and technical change, etc.) and endogenous factors (domestic policies and welfare cutbacks) affect the income distribution.

6.4 Concept and Need of Income Redistribution

Some degree of inequality may be an incentive for the growth of economies, as some concentration of income is a prerequisite for enhancing the rate of investment, if, and only if, the marginal propension to invest by the

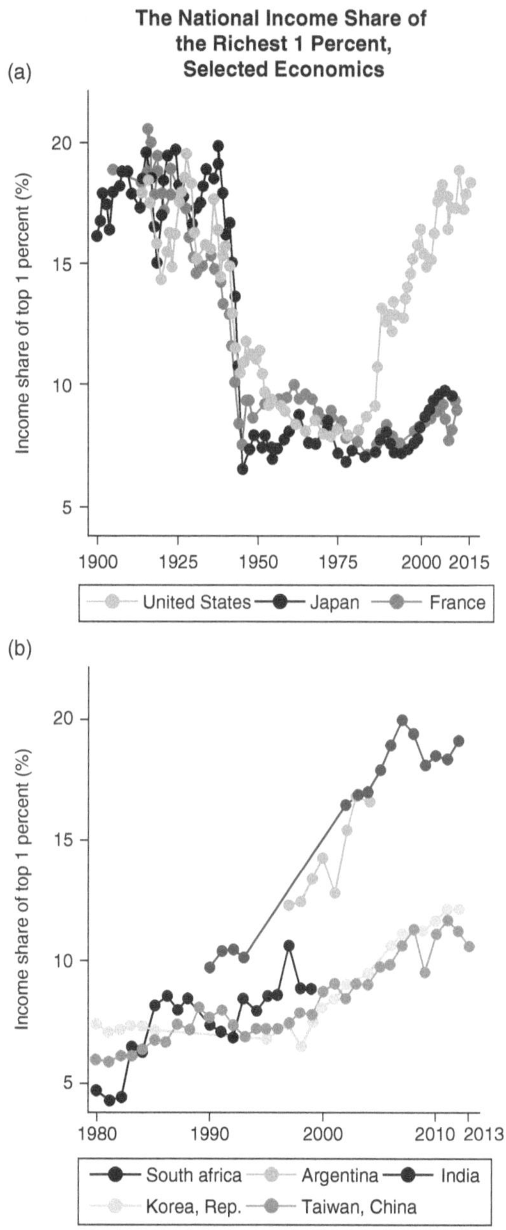

Figure 6.2 Trends of income inequality in industrialized and developing countries. (a) Industrialized economies since 1900. (b) Developing economies since 1980. *Source:* Adapted from calculations based on data from WID (World Wealth and Income Database) https://www.wider.unu.edu/project/wiid-world-income-inequality-database.

wealthier is high enough (according to Keynesian economics). They could instead indulge in "conspicuous consumption" (Veblen 2005).

Although some degree of inequality is necessary for economies to grow, a high and sustained unequal distribution of income, or the lack of access to opportunities for large segments of the population, can generate great social costs at the micro level and may lead to damaging macro-level consequences in the form of political and social instability, which hamper economic productivity. If inequality is excessive, there is a need for income redistribution at the micro and macro levels.

Income redistribution transfers income from wealthier to poorer individuals through some social reallocation mechanism, such as monetary policies, progressive taxation, welfare payments, charity, and public services. There are two schools of thought regarding the effects of income redistribution. Some argue that it often involves policies that distort behavior that may lead to reduced social welfare, focusing on narrow efficiency criteria. Their proponents argue that it helps to resolve social problems, reduce excessive inequality, and promote the development of society. Still, their opponents declare this practice to be a form of theft because it takes control of some earners' funds to give to others. Given the importance of income redistribution for the welfare of individuals, there has been a substantial debate between policymakers and economists over the necessary extent of income redistribution in recent decades.

6.5 Different Redistribution Methods

6.5.1 Cash Transfers

Cash transfers refer to direct payments from a public authority to poor people. Cash transfers tend to give recipients the greatest utility increase compared to any other government program of equal cost, as they only induce income and have no substitution effects. Cash transfers also incur lower administrative costs. In several Organization for Economic Cooperation and Development (OECD) countries, cash transfers have reduced income inequality more than taxes. For instance, on average, 75% of reduction in inequality between market and disposable income is due to transfers, and the rest is due to taxes (Joumardet al. 2012). In Figure 6.3, cash transfer shifts the budget curve (BC) to the right, showing more income to consume on two goods.

In-kind transfers include vouchers for the provision of public goods and tend to enhance individuals' welfare in terms of increased consumption of

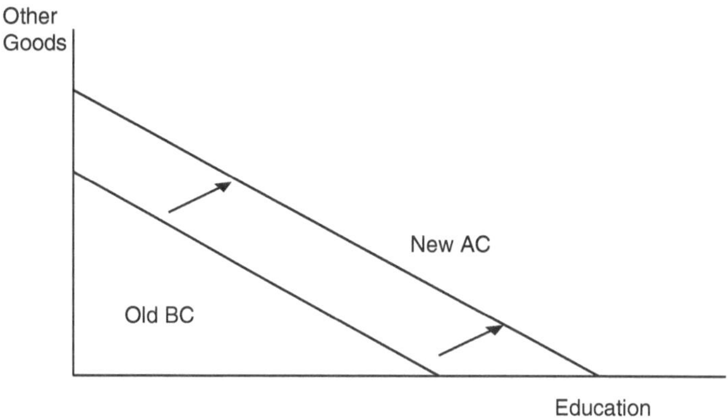

Figure 6.3 Cash transfer. *Note:* *BC refers to the line budget curve.

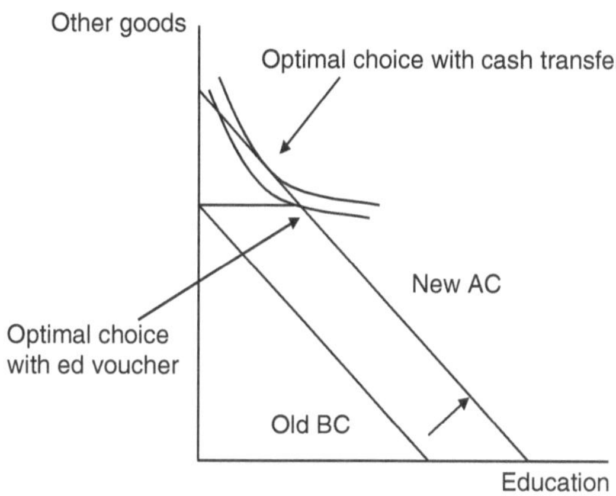

Figure 6.4 In-kind transfer. *Note:* *BC refers to the line budget curve.

targeted goods. They are easier to target than cash payments (Figure 6.4; Pinto 2004).

Figure 6.5 compares the subsidy and cash transfer redistribution methods. In the case of cash transfer choice, the consumer enjoys more cash income to spend on the two goods, and his utility is more than the subsidy. In the case of subsidy, the consumer is restricted to using only one good, e.g. education vouchers, and is on the lower indifference curve.

In this tax system, higher-income people pay higher taxes than low-income people. Hence, the tax amount varies with the percentage of income of earners. Progressive income tax is targeted to transfer the earned revenue from tax to the low segment of the society through a redistribution mechanism. This way, the income inequalities between different income classes may be reduced. In

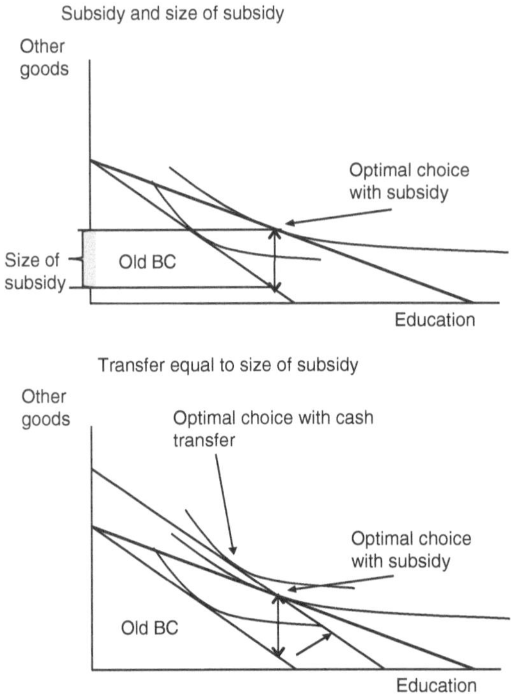

Figure 6.5 Comparing different redistribution methods. *Note:* *BC refers to the line budget curve.

addition to the significant role of taxation in reducing income inequality, it also generates revenues available for investing in public goods and allows transfers that ultimately enhance the welfare of the poorer section of the economy. However, its drawback is that people tend to reduce their taxable income by either working fewer hours (productivity effect) or reporting less income earned (avoidance effect). There is also the black-market effect, where people employ workers "in cash" without declaring them—resulting in job insecurity, unfair wages, poor safety precautions, no medical cover, etc. These underground black markets may have serious implications in underestimating the income inequality figures, leading to misleading policy formulation.

Redistribution of income involves costs, but they are less than the overall benefits to society (Galasso 2014). These benefits are due to the diminishing marginal utility of money. It means that an additional unit of money is worth more for people with less money. For example, someone owning luxurious cars like a Mercedes or an Audi has little interest in buying a Suzuki. Still, the individual who does not own a car and has commuter problems will see great value in owning a Suzuki car.

There is a dire need for time to examine the effects of taxation on income redistribution, keeping in mind the trade-off between equity and economic growth. One must look at the overall effects of any tax reform on the fiscal regime and not just whether taxes are progressive or regressive. This is because disposable income depends on the tax system and benefits. Most developed nations have well-functioning tax systems that generate handsome amounts of tax revenue, which is available for achieving a significant amount of income redistribution.

6.5.2 Negative Income Tax

Another important method of income redistribution is through a negative income tax (NIT). Although it tends to reduce the incentive to work, its welfare effects overcome this negative externality. Poor workers, or people unable to work, receive some financial support. It also simplifies the administration of a welfare system as it is included in the federal tax system. Its drawback is that it is very expensive to provide incentives to poor workers or disabled people. Although it enhances the welfare of workers by giving incentives in the context that tax is negative, workers tend to shorten their working hours.

In Figure 6.6, NITs increase nonlabor income and reduce the net wage. If we treat leisure as a normal good, then an increase in nonlabor income and a reduction in net wage cause the labor supply to reduce. Different tax scenarios have been shown; Point A is the initial equilibrium point. An individual chooses the NIT program and bundle Cover A. NIT has the disincentive effect of doing less work, while the working poor receive some income support through reduced taxes. But NIT is very costly if it provides a reasonable incentive to work and, at the same time, covers the basic needs of those unable to work. It also requires coordination between federal and provincial governments.

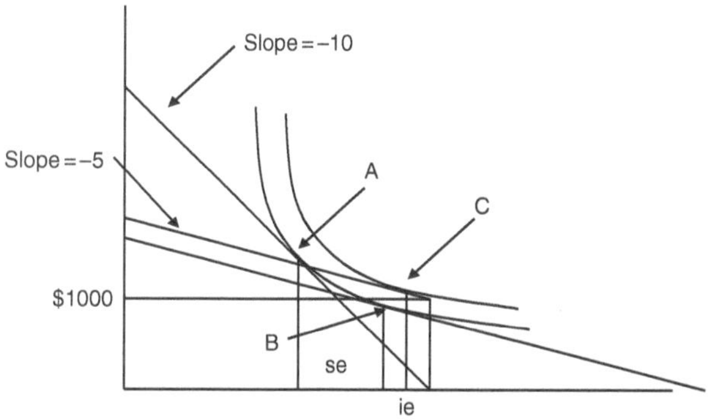

Figure 6.6 Comparing different tax scenarios.

6.6 Case Studies of Developing Countries

6.6.1 Cash Transfers and Redistribution in Latin America

In Latin America, cash transfers have gained significant importance since the mid-1990s as an effective social intervention. The basic objectives were to protect labor markets in the region. They provided cash to low-income families with children on predefined conditions, such as school enrollment and compulsory participation by women in health training. This was the means of breaking a vicious cycle of poverty by developing human capital. Empirical literature has documented that spending money on children of poor households has long-term effects due to increased human capital investment in Latin America. Since the mid-1990s, the expansion of these interventions on human capital has implied that now nearly 20% of the population live in households covered by these transfers (ECLAC 2016). Table 6.1 provides the effect of the cash transfer in terms of budget on poverty and redistribution effects in eight Latin American countries. All countries assigned almost 0.5% of gross domestic product (GDP) to these transfers except for Ecuador, where the figure reaches 1.12% of GDP.

Figure 6.7 shows the annual transfers, which are different across different countries. It indicated that the per capita annual transfer in Uruguay is four times that in Peru and more than 10 times that in Bolivia, depicting the heterogeneity of these interventions.

Although the empirical literature has called it a unique model, it covered and benefited fewer people than claimed in practice. It helped to enhance

Table 6.1 Coverage and budget of selected child transfer programs in Latin America.

Country	Program	Budget (% GDP)	Year
Bolivia	Bono Juancito Pinto	0.23	2011
Chile	Chile Solidario	0.09	2011
Costa Rica	Avancemos	0.20	2013
Ecuador	Bono de Desarrollo Humano	1.12	2013
Panama	Red de Oportunidades	0.14	2011
Paraguay	Tekopora	0.14	2013
Peru	Juntos	0.17	2013
Uruguay	Asignaciones Familiares	0.37	2013

Source: Adapted from ECLAC (2016).

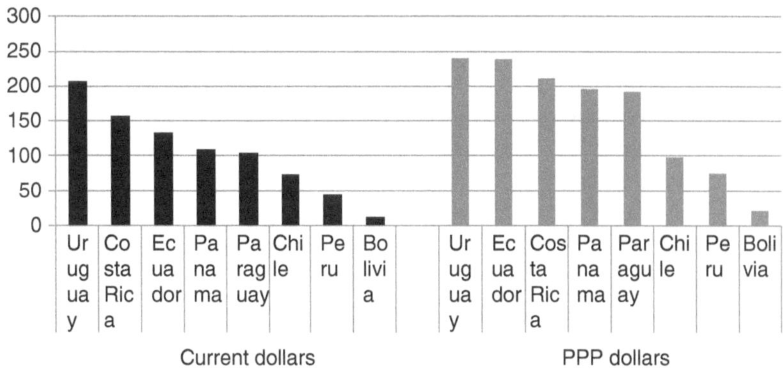

Figure 6.7 Annual cash transfer (Ct) per person for different countries. *Source:* Amarante and Brun (2016)/UNU-WIDER.

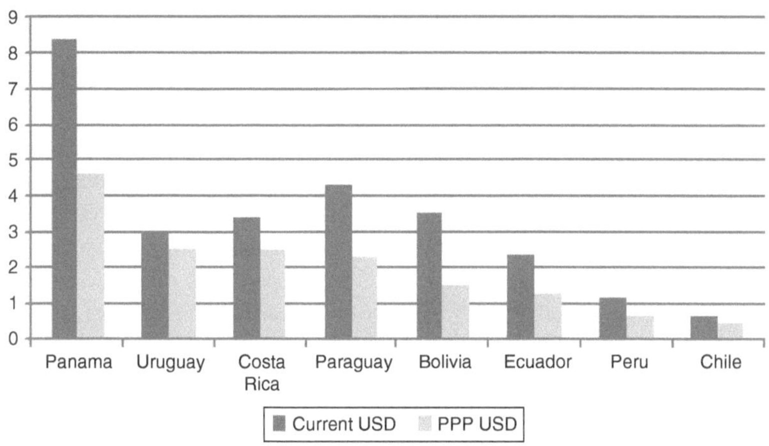

Figure 6.8 Reduction in income inequality by dollar invested. *Source:* Amarante and Brun (2016) and ECLAC (2016)

welfare for some poor people, but income distribution was not significantly affected. The beneficiaries in the region remained highly vulnerable despite the cash transfers (Figure 6.8; Table 6.1).

6.6.2 Redistribution in OECD Countries

The GDP and employment rate in OECD countries started growing in 2010. The economic recovery caused improvement in household income and labor markets. However, this economic recovery did not cause inclusive growth and a reduction in income inequality. Therefore, the income inequality rate remained higher in the last decade despite improving employment rates and declining unemployment. Figure 6.9 shows that the average Gini coefficient of disposable household income value was 0.318 in 2013/2014, which was only marginally higher than in 2007 but had the highest value in the mid-1980s. In the case of Turkey, income inequality reduced by two points in 2010, which reflects the improvement in labor income but increased significantly in Estonia. Substantial changes were less noticeable in all other countries.

Redistribution has affected income inequality. Inequality and changes in it were more pronounced among the working-age population than the total population. Market income inequality (labor and capital income plus private transfers) among the working-age population remained higher despite economic recovery since 2008, as shown in Figure 6.10.

Figure 6.10 presents OECD countries' income inequality before and after redistribution through transfers and taxes.

Redistribution through income taxes and cash transfers reduced income inequality by about 27% in OECD countries. Most redistribution effects (around two-thirds) are due to cash transfers, as shown in the figure's distance between the market income and public cash transfers lines. The rest of the third account is due to taxes. In 2013, the disposable income level among the working-age population was back to the precrisis level despite a deficit in market income. This lower redistribution posed a challenge for policymakers. Widening income inequality between rich and poor has raised awareness about the need to restore growth and equal benefits to all members of society.

6.7 Linkage Between Income Distribution and Food Security

Food security is a multifaceted concept that directly impacts human productivity and development. According to the World Food Summit in 1996, *Food security exists when all people, at all times, have physical and economic access to sufficient, safe and nutritious food that meets their dietary needs and food preferences for an active andhealthy life.*

Food insecurity is due to several reasons, such as liquidity constraints, low agricultural productivity, water shortages, climate change, and political instability. The aforementioned limiting factors are not the only ones: income inequality is one of the major causes of food insecurity.

In addition to the importance of food for a healthy and productive human life, its significance can also be revealed because it accounts for a substantial

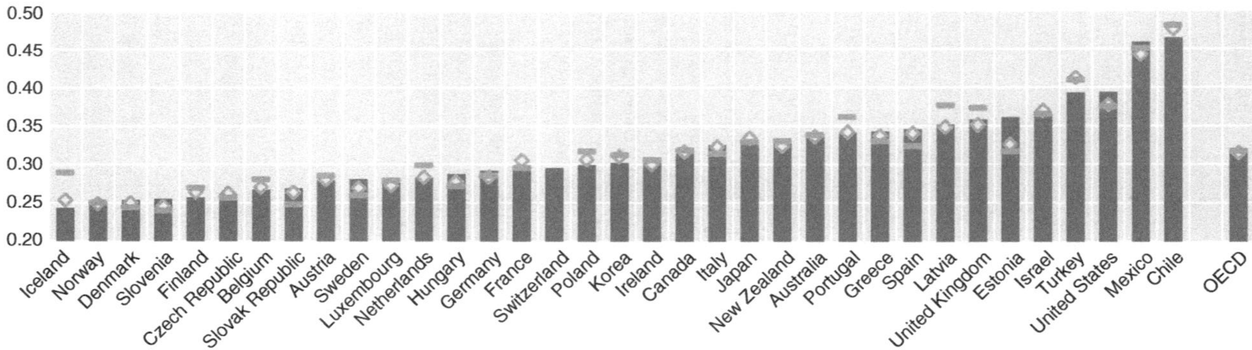

Figure 6.9 Income inequality across the globe.

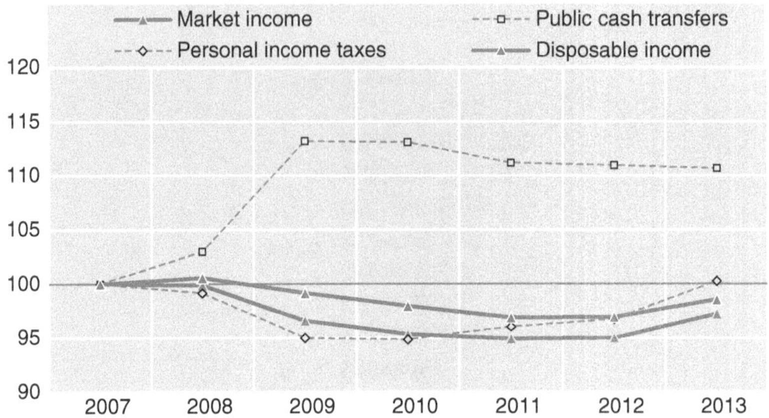

Figure 6.10 Income inequality before and after redistribution through taxes and transfers.

part of a typical household budget. The economic development of a society depends on its endowment of human and nonhuman resources. The productive capacity of human resources depends on how they are fed and medically cared for. Hence, understanding the relationship between income inequality and food security is one of the major challenges for economic and social development in the contemporary world. Income inequalities generate structural problems in an economy that hinder economic growth and generate poverty, social instability, and food insecurity.

In recent times, the body of literature has shifted from the availability of food in an adequate supply to the distribution of food (Barrett and Lentz 2010). The research is reoriented, and there is great attention on food insecurity in rich nations with a disproportionate amount of the world's available per capita food. A significant number of food-insecure people in wealthy nations face difficulty getting access to healthy and nutritious food to live a healthy life.

The relationships between income distribution and food security are complex and imprecise. The rapidly growing population and uneven economic and social development between and within regions make the relationship between income inequality and food security quite complex. This relationship is clearer in the case of the poor than in wealthier nations. Inequality is not only determined in terms of the income distribution but also considers the access to nutrition, health, and social services which are prerequisite in explaining that only economic growth cannot reduce food insecurity, malnutrition, and poverty level. Income inequality causes undernutrition and overweight/obesity because the high cost of nutritious food makes the poor shift toward cheap and nutrient-deficient foods (WHO 2019).

Moreover, inequalities within households help explain why economic growth that tends to reduce extreme poverty may not necessarily affect malnutrition and food insecurity. This phenomenon occurs not only during the economic boom but in all periods. So, an economic slowdown in countries with a high degree of inequality has a disproportionately negative effect on food security.

There is a substantial variation in the level of food security and income inequality between countries due to differences in economic development and a country's economic model. This issue is more prominent in less wealthy economies and the poorer sections of society in developed economies.

The mainstream economic theories cannot satisfactorily describe the issue of food security and income inequality in the face of economic globalization. However, there is a dominant view that reducing income inequality will improve overall social welfare (Thewissen et al. 2015). There is a strong relationship between food security, poverty, and income inequality. Economists have focused intensively on the issues of income inequality, poverty, and their linkages with food security in the last decades of the twentieth century. In this regard, a significant contribution to the linkages between food insecurity and income disparities was made by Sen (2002), who emphasized the importance of purchasing power in eliminating poverty, improving food security, and enhancing people's role in decision-making to improve the well-being of the society. Rosen and Shapouri (2001) also connected food security in less developed countries to people with low-income purchasing power. Similarly, Agwu and Oteh (2014) stressed the reduction of income disparities to improve Nigeria's food security. Popławski and Rutkowska-Podolowska (2011) have emphasized the need for sustainable development by reducing income inequality and improving food security for the poorer sections of society.

Food insecurity is also linked to other social, economic, and demographic factors, such as level of education, age of household head, and distance of villages to big cities and food markets. For instance, Magña-Legmus et al. (2016) have reported that education and demographic factors play a key role in shaping food security and income stability in Mexico. De Castro and Di Mambro (2013) paid attention to the inadequate distribution of food and its waste.

To reduce inequality, there should be effective redistribution and shift toward more inclusive growth patterns that increase poor households' incomes faster than the normal rate. Efficient and effective redistribution measures can play a significant role in equalizing income and opportunities.

6.8 Conclusion

Income inequality is the burning issue of the time, and causing income inequality is a defining issue of our time. It is also a cause of embedded uncertainty and vulnerability (World Bank 2018). A country is said to have income inequality when every member of the population does not have the same divide in income. Although significant efforts are made to reduce extreme poverty globally, income inequality is still high. Most of the efforts in reducing poverty have come through increased income, but not through reductions in income inequality.

References

Agwu, M.N. and Oteh, U.O. (2014). Analysis of income inequalities and food security among farmers in Abia State, South Eastern Nigeria. *Scientific Papers Series Management, Economic Engineering in Agriculture and Rural Development* 14 (3): 7–23.

Amarante, V. and Brun, M. (2016). Cash transfers in Latin America: effects on poverty and redistribution (No. 2016/136). WIDER Working Paper.

Assa, J. (2012). Inequality and growth re-examined. *Technology and Investment 3* (01): 1.

Atkinson, A.B. (2015). *Inequality: What can be done?* Harvard University Press.

Babu, M.S., Bhaskaran, V., and Venkatesh, M. (2016). Does inequality hamper long run growth? Evidence from emerging conomies. *Economic Analysis and Policy* 52: 99–113.

Barrett, C.B. and Lentz, E.C. (2010). Food insecurity. In: *The International Studies Encyclopedia*, vol. IV (ed. R.A. Denmark). Chichester, UK: Wiley-Blackwell.

Barro, R.J. (2000). Inequality and growth in a panel of countries. *Journal of Economic Growth* 5 (1): 5–32.

Baulch, B. (ed.) (2011). *Why Poverty Persists: Poverty Dynamics in Asia and Africa*. Edward Elgar Publishing.

Berg, A.G. and Ostry, J.D. (2011). Equality and efficiency:is there a trade-off between the two, or do they go hand in hand? *Finance & Development* 48 (003): 12–15.

Bourguignon, F. and Dessus, S. (2009). Equity and development: political economy considerations. *No Growth without Equity?* 45: 45–69.

Carvalho, L. and Rezai, A. (2015). Personal income inequality and aggregate demand. *Cambridge Journal of Economics* 40 (2): 491–505.

Clements, M.B.J., de Mooij, R.A., Gupta, M.S., and Keen, M.M. (2015). *Inequality and Fiscal Policy*. International Monetary Fund.

Dartanto, T. and Nurkholis (2013). The determinants of poverty dynamics in Indonesia: evidence from panel data. *Bulletin of Indonesian Economic Studies* 49 (1): 61–84.

De Castro, P. and Di Mambro, A. (2013). The European CAP Reform and the Mediterranean area, CIHEAM-Watch Letter 27 [online]. http://www.ciheam.org/images/CIHEAM/PDFs/Publications/LV/WL27.pdf (accessed 30 December 2016).

Easterlin, R.A. (1995). Will raising the incomes of all increase the happiness of all? *Journal of Economic Behavior & Organization* 27 (1): 35–47.

ECLAC (2016). Non-contributive social programs database. http://dds.cepal.org/bpsnc/ (accessed 10 November 2016).

Galasso, V.N. (2014). *The Drivers of Economic Inequality: A Backgrounder for Oxfam's Inequality Campaign*. Oxfam America Research Backgrounder Series.

Joumard, I., Pisu, M., and Bloch, D. (2012). Tackling income inequality: the role of taxes and transfers. *OECD Journal. Economic Studies* 2012 (1): 37.

Kendall, M. and Stuart, A. (1979). *The Advanced Theory of Statistics: Inference and Relationship*. London: Hodder Arnold.

Krantz, L. (2001). *The sustainable livelihood approach to poverty reduction. SIDA. Division for Policy and Socio-Economic Analysis*, vol. 44, 1–38.

Krishna, A. (2007). For reducing poverty faster: target reasons before people. *World Development* 35 (11): 1947–1960.

Kuznets, S. (1955). Economic growth and income inequality. *The American Economic Review* 45 (1): 1–28.

Magna-Legmus, D., Ishdorj, A., Rosson, P., and Lara-Alvarez, J. (2016). Determinants of household food insecurity in Mexico. *Agricultural and Food Economy* 4 (10): 2–19.

McKay, A. (2002). Defining and measuring inequality. Briefing Paper No. 1 (1 of 3).

Nino, F. S. (2015). Sustainable development goals—United Nations. *UN Sustainable Development* http://www.un.org/sustainabledevelopment/sustainable-development-goals/ (accessed 1 May 2016).

Obst, T. (2015). Long-term trends in income distribution: a global perspective (No. 29). Global Labour University Working Paper.

Ostry, J.D. and Berg, A. (2011). *Inequality and Unsustainable Growth; Two Sides of the Same Coin? (No. 11/08)*. International Monetary Fund.

Ostry, J., Berg, A., and Tsangarides, C. (2014). Redistribution, Inequality, and Growth. International Monetary Fund.

Pinto, S.M. (2004). Assistance to poor households when income is not observed: targeted in-kind and in-cash transfers. *Journal of Urban Economics* 56 (3): 536–553.

Poplawski, L., andRutkowska-Podolowska, M.(2011). Social economy and sustainable development. *Non-Conferenced Review Proceedings as part of project solutions, "VEGA, KEGA, APVV, ENER SUPPLY."* Bratyslawa, 99–103.

Rajan, R.G. (2010). *Fault Lines: How Hidden Fractures Still Threaten the World Economy*. Princeton, New Jersey: Princeton University Press.

Rosen, S. and Shapouri, S. (2001). Effects of income distribution on food security. *US Department of Agriculture Information Bulletin , (765–2) p. 2.*

Sen, A.K. (2002). Globalization, inequality and global protest. *Development* 45 (2): 11–16.

Singh, K. (2009). *Rural Development: Principles, Policies and Management.* Sage Publications.

Spielman, D.J., Malik, S.J., Dorosh, P., and Ahmad, N. (Eds.). (2017). *Agriculture and the Rural Economy in Pakistan: Issues, Outlooks, and Policy Priorities*. University of Pennsylvania Press.

Thewissen, S., Kenworthy, L., Nolan, B., Roser, M., andSmeeding, T. (2015). Rising income inequality and living standards in OECD countries: How does the middle fare? (No. 656). LIS Working Paper Series.

Veblen, T. (2005). *Conspicuous consumption*, vol. 38. UK: Penguin.

World Bank (2018). *Overcoming Poverty and Inequality in South Africa: An Assessment of Drivers, Constraints and Opportunities.* World Bank.

World Health Organization (2019). *The State of Food Security and Nutrition in the World 2019: Safeguarding Against Economic Slowdowns and Downturns*, vol. 2019. Food & Agriculture Organization.

7

Market Availability

Hammad Badar, Burhan Ahmad, and Muhammad Hafeez

Institute of Business Management Sciences, University of Agriculture, Faisalabad, Pakistan

7.1 Introduction

Ensuring food security for the world population has recently emerged as the most pressing problem. The imperative to double global food production for more than nine billion people expected on earth in 2050 has captured the imagination of global leaders, politicians, policymakers, academia, and industry (Randell et al. 2022; Wang 2022). Across the globe, public and private sector organizations are endeavoring to develop policies, approaches, and action plans for ensuring short- and long-term balance between food production and consumption (Chávez-Dulanto et al. 2021; IFPRI 2018). Producers are being encouraged to enhance and diversify their production to meet market demands. Increasing efforts are underway to link farmers with markets.

Within the complex food system, achieving food security is not an easy task (Timmer 2014). Food security can be ensured when all stakeholders such as farm suppliers, growers, and traders achieve a minimum desired profitability level. At the same time, consumers have access to food that is of good quality and safe to consume. The Food and Agriculture Organization (FAO) defined food security at the World Food Summit in 1996 as "a condition in which all people, at all times, have physical and economic access to sufficient, safe and nutritious food to meet their dietary needs and food preferences for an active and healthy life." This concept of food security has revealed four widely accepted underlying

Food Security in the Developing World, First Edition. Edited by Muhammad Khalid Bashir, Steven G.M. Schilizzi, and Ghaffar Ali.

dimensions, i.e. food availability, access, utilization, and stability (Chávez-Dulanto et al. 2021).

Food availability is concerned with ensuring an adequate supply of appropriate quality food either through domestic production or imports. Food access refers to whether individuals can access and afford food to sustain their lives. Utilization highlights the importance of nonfood inputs such as water, sanitation, and health care in ensuring that all physiological needs of human beings are met. Stability implies the ability of the food system to ensure that the population always has access to adequate quantity and quality food (Alabi and Ngwenyama 2023).

Food accessibility refers to both economic and physical access. Households can access food in two ways. They can either grow their food for consumption or purchase it from the market. Consumer food requirements have expanded and diversified due to lifestyle changes and a rise in per-capita incomes. They may grow their food in smaller quantities from their resources, which is generally insufficient to meet all their dietary needs. Consequently, they must resort to markets for purchasing their food requirements.

For a long time, the concept of food security was limited to food production, and all efforts in the last century were directed to increase production (Lencucha et al. 2020). Largely, this approach helped in meeting the food demand of the world population (Chávez-Dulanto et al. 2021). However, a rapid rise in population and changing consumer preferences have raised questions about the effectiveness of this approach in the future. There is a growing realization that paradigm changes in the agricultural system are needed because production enhancement alone cannot ensure food security. There is a need to shift producers, mainly smallholders, from traditional to more innovative farming and link them to markets in various ways, such as by providing them with market information, promoting contract farming, and organizing farmers into associations, groups, or cooperatives (Njuki et al. 2011).

Nowadays, all agricultural and rural development strategies embrace market development and try to link farmers with markets. It is widely believed that well-functioning markets are essential for ensuring overall food security and welfare gains to all stakeholders including producers and consumers. Therefore, making markets work for farmers, especially smallholders, is increasingly included in policy priorities (Njuki et al. 2011). The chapter focuses on how markets work and what their structure is? In terms of market availability, what options do producers and consumers have? What are the market-related issues that prevent achieving food security, and how can the role of the markets be improved?

7.2 Markets and Their Functions

Markets play an essential role in countries' socioeconomic development by significantly impacting the lives of the people directly and indirectly linked with them, either as suppliers, buyers, traders, or workers. They generate opportunities to earn income and contribute to reducing poverty and hunger in developing countries. Markets are institutions or mechanisms that link sellers and buyers with each other and enable them to trade agricultural products according to their requirements. In this way, they facilitate the exchange of goods and services which are essential for ensuring the food security of a country.

In general, a market is considered a place where buying and selling of commodities take place and indicates the aggregate demand and supply of commodities in any area or place. In the technical terms of business management sciences, it refers to an exchange process along with a price-setting mechanism (Mohy-ud-Din and Badar 2011). Markets can be classified in many ways, and some are as below.

a) Location: primary, secondary/wholesale, retail, and terminal markets
b) Coverage: local, national, and international markets
c) Regulation: free and regulated markets
d) Competition: highly competitive and monopolistic markets
e) End users: consumer and business/industrial markets
f) Products: input and product markets

Markets perform numerous crucial functions such as the distribution of agricultural inputs and outputs, transformation of market information and risks, conversion of raw agricultural products into value-added products, and allocation of resources to produce agricultural commodities demanded in the markets. Some of the important functions are presented below.

7.2.1 Exchange Function

The foremost function of a market is to accomplish the exchange process between sellers and buyers. The spatial separation of production and consumption of agricultural commodities further increases the importance of this function along with the physical activities such as transport, distribution, and storage performed in the markets. Agricultural commodities are produced in peri-urban and distant rural areas. In contrast, their consumption takes place mostly in urban areas. Neither can producers reach out to consumers for selling their produce nor can consumers frequently visit producers for buying their food. The market mechanism bridges the distance gap and transfers commodities from production to consumption places. The wholesale markets serve as a hub where scattered agricultural produce from many villages first converges and then diverges to widely spread consumers through a network of traders and transportation and storage facilities. In this way, the wholesale markets play a key role in developing horizontal and vertical coordination in the agricultural marketing system.

7.2.2 Price Determination

An important function of agricultural markets is the price determination of agricultural commodities. In agricultural markets, prices are determined based on the daily demand and supply of agricultural produce. Markets provide an environment where demand and supply can meet in some balance to determine wholesale prices. Considerable fluctuations often take place in the demand and supply of agricultural commodities, and markets respond through price adjustments. Additionally, prices for the same product may vary not only daily over time but also locally across regions and localities. As a result, there can be multiple prices at the same time.

7.2.3 Resource Allocation

Well-functioning markets enable the accomplishment of mutually beneficial exchanges between sellers and buyers. In doing so, the interaction of demand and supply forces generates market signals through price determination. Higher prices indicate market demand and signal producers to grow more of that crop to earn profit. Accordingly, producers allocate more resources to the crops which are likely to be priced high in the markets. When markets distort price signals, producers do not correctly interpret them. They would misallocate resources by producing crops that are not demanded in the market and hence would result in the wastage of resources.

7.2.4 Policy Implementation

Markets have become an important policy instrument for smallholder inclusion, increased food security, and poverty reduction. Macro-level policies for improvement can only be effective when markets function well and transmit signals to all stakeholders in the system (Barreti and Mutambatsere 2008). Nontransmission of price incentives generates gaps between what producers supply and what consumers demand. Only well-functioning markets can bridge this gap. However, market inefficiencies and failures may sometimes constrain achieving desired policy objectives. Therefore, the critical role of markets in successfully implementing policies cannot be overlooked.

7.2.5 Value Addition

Markets also play an important role in value addition by creating form, time, place, and possession utilities (Mohy-ud-Din and Badar 2011). The processing industry transforms raw agricultural commodities into consumable forms desired by consumers such as jams, juices, and squashes. This prevents seasonality-related shortages and surpluses because processed agricultural products are consumed all year long. Storage and warehousing facilities overcome time lag and thus help in agricultural commodities' regular consumption and distribution. Often agricultural commodities are graded and packed in markets which further add to the value of these commodities.

7.2.6 Employment Generation

Markets generate sizeable employment opportunities. In developing countries, market operations are mostly carried out manually. Numerous people earn their livelihoods in markets as merchants, commission agents, brokers, export and import handling agents, laborer, and transporters. A wide range of skilled, semi-skilled, and unskilled workers earn their income by performing various functions such as buying, selling, transportation, storage, loading and unloading, cleaning, sorting, and grading. The public sector employees engaged in market management operations also earn their livelihood by performing various functions such as supervision, security, fee collection, and infrastructure maintenance.

7.3 Stakeholders in Agricultural Markets

In agricultural markets, intermediaries and growers operate to earn their livelihood. In addition, consumers and policymakers also have stakes in agricultural markets. Broadly, these stakeholders can be classified into five categories. They include producers, consumers, traders, trade supporters, and trade planners/decision-makers. Each of these stakeholders has its own specific goals, which may or may not be consistent with each other.

7.3.1 Producers

Producers or farmers are the main stakeholders in agricultural markets. What they grow is traded in agricultural markets. They produce a wide range of agricultural commodities for self-consumption and selling in markets for earning profit. Some perform harvest, postharvest, and marketing activities themselves and bring their produce to markets to capture greater profits. At the same time, many producers avoid performing these activities due to a lethargic attitude, credit constraints, market risks, etc. Market intermediaries perform these activities in such cases and earn the income forgone by producers. Producers' primary goal is to maximize their income and avoid various types of risks involved in the production and marketing of their crops. They also want developed and well-functioning markets that can ensure improved market position and higher and stable prices of their produce.

7.3.2 Consumers

Although consumers are not direct participants in agricultural value chains, their influence is considered a derived function which is executed through traders. Consumers need a wide range of food products, such as cereals, meat, dairy products, and pulses, to satisfy their needs. They cannot grow all of them domestically and must resort to markets for buying food products. Over time, consumer requirements have not only changed but also expanded. When buying products in markets, they look for good quality products that are safe to consume and available at lower prices, along with better marketing services. For profitability, businesses operating along the food value chains must work together and align their practices with consumer requirements.

7.3.3 Traders

Traders operating in markets play a critical role in relating producers with consumers. Along the value chains, numerous traders such as village dealers, commission agents, brokers, wholesalers, and retailers operate to facilitate product flows toward consumers. Mostly, they render valuable services in the marketing system and add form, time, place, and possession utilities. However, some of them increase the price spread without adding value to the system. Gaining profit is the principal objective of all traders, which they try to achieve through high volumes, time, and cost-efficient purchasing. They also want minimal market risks and no trade restrictions in markets.

7.3.4 Trade Supporters

Trade supporters are not direct participants in the trading activities of the market. Instead, they are interested in improving marketing efficiency and smooth exchange of goods in markets. Hence, they perform various activities that support, facilitate, and smoothen market trading activities. Trade supporters may include grading and standardization bodies, lending institutions, information service providers, and research and development organizations. The main goals of trade supporters are the smooth functioning of markets and general support for market exchange functions such as infrastructure and communication.

7.3.5 Trade Planners/Decision-Makers

Trader planners or decision-makers are responsible for decision-making about markets. Their main goal is to enhance the welfare of all stakeholders in the marketing system by ensuring adequate returns to producers and good quality food products to consumers at fair prices. At the aggregate level, their goal is to frame policies that boost agricultural growth and exports so that the agriculture sector can contribute to its full potential to the socioeconomic development of their country. Sometimes, their political and social goals lead them to frame policies and decisions that may not yield equitable or fair returns to all.

7.4 Agricultural Marketing Systems in Developing Countries

Producers and consumers are the two main stakeholders in the agricultural marketing system. Food produced by producers flows through the system and reaches consumers for ultimate consumption (Figure 7.1). Producers have a choice to produce a wide range of agricultural commodities which depends on various market and nonmarket factors. Market factors mainly include prices, terms, and seasonality. Higher market prices provide a signal to producers to grow specific crops. Similarly, they prefer to grow those crops with favorable terms regarding cash or spot payments and seasonable suitability.

The nonmarket factors include technology, inputs, and alternate resource use. Every crop has its specific production technology, and crop productivity heavily depends on its efficient and effective use. Producers tend to produce those crops with the production technology they are familiar with.

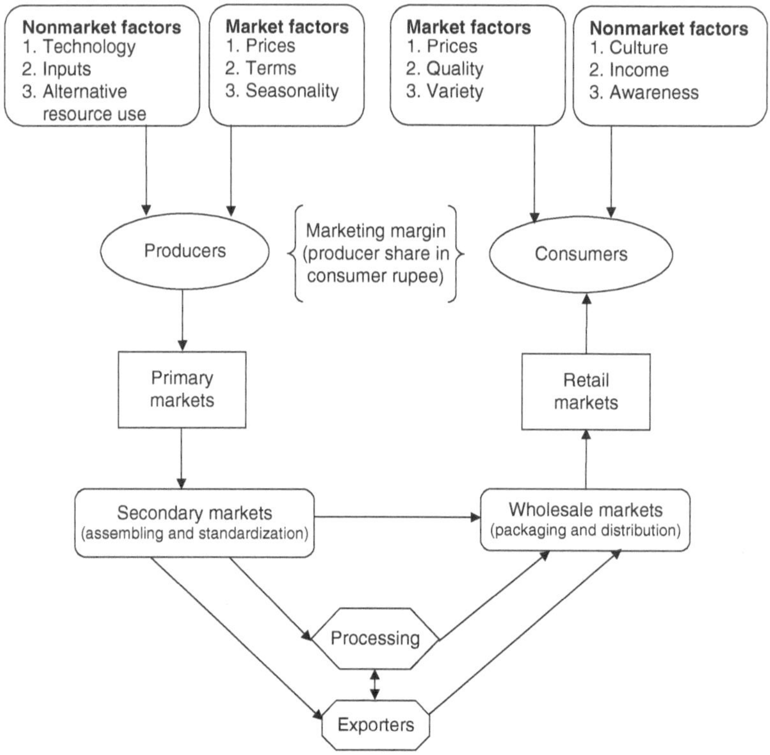

Figure 7.1 Typical structure of an agricultural marketing system. *Source:* Adapted from Mohy-ud-Din and Badar (2011).

Availability and reliability of inputs also play a critical role in producers' choice of crops. Every crop has a specific input requirement for yielding optimum productivity. However, producers may often be constrained in applying adequate amounts of input due to financial or technical issues. In such cases, they avoid producing those crops for which they cannot ensure adequate and timely availability of inputs. Producers also consider alternate uses of their resources such as land, labor, and water in choosing various crops.

In developing countries, agricultural products pass through several hands before reaching consumers. First, most of the producers sell their produce to village dealers who are located near their fields. Then, village dealers collect agricultural produce from several producers for selling in secondary markets where scattered agricultural produce is assembled and standardized. In the case of raw products, secondary and wholesale markets are the same. Then, agricultural produce is passed on to retailers for final selling to consumers.

Part of the agricultural produce also undergoes processing for value addition. These value-added products then pass through the wholesale and retail markets to reach consumers. Surplus agricultural produce in raw or processed form may be exported. In importing countries, these products move through wholesale and retail markets and reach consumers for ultimate consumption.

As discussed earlier, consumers' requirements have considerably changed and expanded. Nevertheless, various market and nonmarket factors influence them. Market factors include prices, quality, and variety. Price is a critical factor in consumer choice, particularly in developing countries. Consumers prefer buying products that are not only low-priced but also of good quality. They have also become variety seekers because new products rapidly arrive in markets due to competition among food producers and marketers. They want products that fit their lifestyles.

Consumer food preferences are also affected by nonmarket factors, including culture, income, and awareness. Culture is a way of life and the main guiding force for what and what not to consume. Consumers follow their culture in their purchase decisions. For instance, the consumption of wine and non-halal food is prohibited in Muslim culture, so such products are not traded in Muslim societies. Income is a significant determinant of demand as it affects the affordability of consumers. Generally, low-income consumers avoid buying products that they cannot afford. Awareness of availability, safety, and technical specification is another important factor influencing consumer choice of food products. Consumers would tend to buy those products about which they have awareness.

7.5 Market Availability Options for Producers

The modern marketing concept does not limit the scope of marketing to only post-production activities. It includes everything from deciding what to produce until the product reaches consumers for final consumption. Therefore, marketing options and alternatives should be carefully examined even before the commencement of production activities. In developing countries, producers have various options for marketing their agricultural commodities. Each option has its advantages and disadvantages, and producers may choose one or more options depending on various considerations such as farm size, production volume and quality, marketable surplus, market linkages and knowledge, and farm location. These marketing options can be identified as direct and indirect marketing and are described below.

7.5.1 Direct Marketing

Direct marketing involves the direct interaction of producers with consumers without the involvement of intermediaries or traders. The farmer's market is an example of direct marketing. In developing countries, the involvement of intermediaries is generally high in agricultural markets. Therefore, greater emphasis is laid on promoting direct marketing, particularly the establishment of farmer markets, to enhance the returns to producers. It is also argued that farmer markets allow consumers to buy good quality fresh produce at reasonable prices directly from farmers. When input costs are increasing, farmer markets offer a unique opportunity to earn a better profit margin for farmers. However, direct marketing entails disadvantages. Producers must spare time for operating farmer markets, which they often cannot afford because of their farm engagements. Extra costs such as transportation and stalls are also incurred, which often producers are constrained to bear due to their all-financial difficulties.

7.5.2 Indirect Marketing

In the case of indirect marketing, agricultural producers interact with intermediaries and sell or hand over their produce to them, who then sell to consumers. The intermediaries may include exporters, commission agents, brokers, processors, cooperatives, retail firms, etc. In developing countries, producers mostly opt for indirect marketing alternatives because of the ease of disposing of their produce, financial constraints, and lack of marketing experience. Mostly, producers sell their produce to buyers who offer higher prices and prompt payments. Sometimes, they are bound to sell to them because of their debt obligations to a particular buyer.

When selling to exporters, processors, or large retail firms, producers may have to consider buyers' volume, quality, grading, packing, and other requirements. The major advantage is that indirect marketing relieves producers of many postharvest and marketing activities and they can focus on their farm operations for the next crop. However, the main disadvantage is profit forgone that producers can capture through direct selling methods. In the following, various indirect marketing options are discussed.

7.5.2.1 Secondary or Terminal Markets

Secondary markets in major cities or large metropolitan areas serve as assembly and distribution centers. They may be termed terminal markets in case products are exported from these markets. Producers transport their produce to intermediaries, mostly commission agents, who then arrange sales through auction or direct sale. After purchasing in bulk, large wholesalers or retail stores generally redistribute the goods in smaller quantities to local markets. Generally, large producers directly send their produce to the terminal market. Mostly, small producers do not send their produce directly to terminal markets because they sell in nearby local markets. Terminal markets generally have relatively better infrastructure and up-to-date market information, enabling producers to interact with buyers. However, they must meet buyers' requirements and deliver products in these markets.

7.5.2.2 Primary or Local Markets

Primary markets are located near villages. Resource-constrained producers who often lack market linkages prefer to sell in these markets. From primary markets, agricultural produce is then transported to secondary markets. By selling locally, producers get money without performing many postharvest and marketing activities. However, they must forego the profit that can be captured by selling in secondary markets. The chance of earning more profit is higher in the secondary markets due to the presence of many buyers who may offer higher prices for their desired quality.

7.5.2.3 Processors/Exporters/Retailers

In areas where processing plants operate, producers get an opportunity to supply them with raw commodities such as milk, fruits, and vegetables and earn good money. In many cases, they contract with processing plants to supply agricultural produce to them. Sometimes, processors provide producers with inputs and training to reduce their unit costs and improve quality and distribution efficiency. Often buying agents or contractors of various exporting or retail firms also operate in production areas for buying agricultural produce directly from producers. Although this option can bring more returns, producers must comply with the stringent process and quality requirements. Often, many fail to meet these requirements and may incur losses.

7.5.2.4 Cooperatives

Cooperatives are voluntary organizations or associations of producers formed to solve common problems mutually. They pool the produce of individual producers with a small marketable surplus for selling in markets. Where cooperatives are functioning successfully, they help producers secure guaranteed markets, higher prices, and lower handling costs. In many cases, they provide various harvesting, postharvest, and marketing services such as grading, packing, storage, cooling, and transportation facilities at operating costs.

7.6 Markets Available to Consumers

Food is a basic need, and consumers buy a wide range of food items, such as cereals, meat, milk, pulses, fruits, and vegetables, to fulfill their dietary requirements. However, due to time and financial constraints, they can neither produce all their food nor visit producers for this purpose. Therefore, they visit markets and buy their food mainly from retailers. In developing countries, many retailers operate to supply consumers with their food requirements. These retailers can broadly be categorized as traditional and modern due to differences in resources and retailing practices. The characteristics and business practices of these two categories are described below.

7.6.1 Traditional Retail Markets

In developing countries, traditional retailers such as street vendors, hand carts, donkey carts, temporary roadside stalls, and permanent shops are predominant. These retailers lack adequate infrastructural facilities due to their weak financial position. The quality of food products sold by these retailers depends on their location. Those in high-income areas sell good quality, while others in low-income areas sell low- to medium-quality items. The business practices of traditional retailers include buying, transportation, grading, cleaning, display, and selling. They procure their supplies from wholesale markets. Consumers belonging to low- and middle-income groups mostly buy from traditional retailers because these retailers are located conveniently near to consumers.

7.6.2 Modern Retail Markets

The presence of modern retailers is gradually growing due to the emergence of supermarkets in developing countries (Chamhuri and Batt 2013; Ghani 2005; Macharia et al. 2013). Modern retailers such as supermarkets have extensive infrastructure and resources for product procurement, storage, display, and sale. Moreover, they sell almost all consumer items, including fresh fruits, under one roof to serve the needs of middle- and high-income consumers. As a result, consumers increasingly prefer these modern retailers. The primary reasons include a rise in per capita income, population growth, urbanization, and changes in consumer tastes and lifestyle (McCullough et al. 2008; Senauer and Venturini 2005; Traill 1997).

7.6.3 Problems of Markets in Developing Countries

Agriculture is a primary sector in most developing countries. Most of the population, particularly those living in rural areas, directly and indirectly, depend on this sector for their livelihood. However, agricultural markets are not well developed and lack modern infrastructure facilities. In addition, several market imperfections and inefficiencies hinder realizing optimal welfare gains for all stakeholders (Barreti and Mutambatsere 2008). The following section discusses some of the problems of markets in developing countries.

7.6.4 Inequitable Returns

Markets play a crucial role in efficiently allocating resources and transmitting information. In doing so, however, they often yield more benefits to resourceful and well-endowed than less resourceful, skilled, and educated stakeholders (Hussain 2018). Van der Meer (2006) points out that food markets worldwide are generally characterized by small-scale producers who lack economies of scale and cannot compete effectively against large-scale farmers on a level playing field in a free market.

Globalization, rise in per capita income, and urbanization have changed consumer preferences for food. However, most small-scale producers in developing countries with limited resources, skills, and competencies cannot fulfill the needs of consumers. Easy and cheap imported food availability further makes it difficult for local growers to sustain themselves in the markets. Comparatively, large-scale producers are better positioned to capture market signals more quickly and align their produce with consumer requirements because of their resources and linkages. They also have access to storage and warehousing facilities which enable them to store their produce until prices get stable or high in markets. In contrast, smallholders lacking this access are compelled to sell their produce even at lower prices.

7.6.5 Predominance of Intermediaries

Markets in developing countries tend to be dominated by middlemen or intermediaries. A variety of market intermediaries, including brokers, commission agents, speculative middlemen, and wholesalers, operate in agricultural markets. Generally, an agricultural commodity passes through more than five hands before reaching ultimate consumers. Producers are often reluctant to bring their produce directly to markets because of a lack of resources and market linkages, which creates room for the entry of intermediaries. Some intermediaries render valuable services such as grading, standardization, and packing. However, many exploit both producers and consumers. Instead of adding value to the system, they are interested in earning their margin by purchasing produce at low prices from producers when supply is high and selling at higher prices to consumers when demand gets high. Often, they enjoy predominance in price determination and selling and buying agricultural produce.

7.6.6 Problem of Transportation

Unlike in developed countries, transportation facilities are poor, inadequate, and inefficient in most developing countries. As a result, transportation of agricultural products and inputs from one place to another often becomes difficult. Many distant areas are not connected to main roads. In the rainy season, many areas become unserviceable, and producers face difficulties in bringing their produce into markets. Inadequate transportation issues also cause postharvest losses. A significant part of the produce is lost or wasted on the way to markets, which not only brings lower returns to producers but also causes a waste of resources and higher prices for consumers.

7.6.7 Inadequate Storage and Warehousing Facilities

Considerable time lag exists between the production and consumption of agricultural produce. Storage facilities are needed to store products until consumers demand them for ultimate consumption. Due to limited resources and inadequate farm-level storage facilities, most producers in developing nations are unable to store their produce until the market conveys signals of higher prices. As a result, they are forced to sell their produce as soon as it is harvested. The lack of storage and warehousing facilities in agricultural markets further reduces producers' ability to bargain and prevents them from receiving fair prices for their output.

7.6.8 Lack of Credit

Producers in developing countries lack credit for carrying out various production and marketing functions. While it takes time to sell the harvested crops, they also need to decide for sowing the subsequent crops, which requires money. Formal agricultural credit facilities are not widely available in rural areas. Where banking facilities are available, high-interest rates, other costs, and cumbersome procedural requirements hinder producers from borrowing from formal sources of agricultural credit. Consequently, they resort to informal credit sources that may exploit them.

7.6.9 Lack of Modern Marketing Knowledge

Producers and other stakeholders operating in agricultural markets lack the know-how and use of modern marketing knowledge and tools. Using promotional activities such as sales promotion advertising and personal selling for agriculture products is not common. Agricultural producers do not know how to brand and pack their produce. As a result, consumers do not get accurate information about agricultural products' prices, quality, and availability. The modern marketing concept revolves around consumers as the focus of all marketing activities. All value chain actors work together to provide what is needed by consumers. However, many value chain actors, including producers in developing countries, are still unaware of the importance of understanding consumer requirements. As a result, they try to push through the chain whatsoever is produced, which often does not fetch reasonable prices in markets because of incompatibility with consumer needs.

7.6.10 Outdated Measuring Equipment

Using old, manual, outdated measuring equipment is still common, particularly in rural areas of developing countries. The use of modern equipment such as computerized and digital scales for measurement is rare. As a result, producers and consumers suffer at the hands of market intermediaries. Due to short weighing, producers receive less money for their produce, whereas consumers must pay more for what they buy.

7.6.11 Lack of Technical Knowledge

Generally, producers and traders in developing countries are not educated and lack technical knowledge about various production and marketing processes of agricultural products. Traditional poor production and marketing practices cause both quantitative and qualitative losses. Part of the produce is often lost or wasted during the production, harvest, postharvest, and marketing phases. Not only do these losses negatively affect returns to producers but also cause a waste of resources. Product quality is also affected, and consumers do not get their desired quality attributes. Food safety issues may also arise due to this lack of technical knowledge of value chain actors who often do not know how various physical, technical, and chemical hazards can impact the lives of consumers.

7.6.12 Lack of Organized Markets

Agricultural markets in developing countries are not well organized, and producers face difficulties in selling their produce directly in markets and earning adequate returns. Producers often avoid bringing their produce directly to markets. Instead, they prefer to sell their produce to local traders at lower prices. Furthermore, markets lack basic infrastructure, and producers may have to wait for reasonable prices for their goods. Intermediaries take advantage of both producers and customers due to their power. The producers' associations are either nonexistent or not working to safeguard the interests of producers.

7.7 Policy Measures for Improved Market Availability

Developing a sustainable and diversified agricultural system capable of reducing food insecurity and improving the livelihood of smallholders is a significant challenge faced by developing countries. In developing countries, agricultural markets are also experiencing a gradual transformation due to technological innovation, changes in consumer preferences and lifestyles, and the emergence of supermarket culture (Birthal et al. 2007). This transformation has generated opportunities and challenges for producers in developing countries. However, they can earn profit only by ensuring consistency and quality in their supplies according to market requirements. Some measures are suggested below to improve market availability for producers and consumers in developing countries.

7.7.1 Entrepreneurial Mindset

In developing countries, there is a dire need to inculcate and foster an entrepreneurial mindset in producers. Currently, producers do not consider their farm a business unit, mainly because they have inherited the land and take it for granted. Furthermore, producers lack innovation and creativity and always endeavor to seek support from public sector institutions for everything from farm supplies to produce marketing. This dependency syndrome is a major hindrance to their way of improvement. An entrepreneurial mindset can only overcome this problem. Producers need to be progressive because the dynamics of agricultural markets have changed. Growing demand for high-value food products has given rise to opportunities, particularly for smallholders who can capture more profit by aligning their production with consumer requirements (Birthal et al. 2007). Only producers with an entrepreneurial mindset can sense and capture market opportunities.

7.7.2 Market Governance

The governance of agricultural markets in developing countries is poor. As a result, efficiency and effectiveness in various market operations are far below the optimal level. Efficient markets cannot be possible without good governance and public policy. For agricultural and rural development, policymakers should prioritize improving market governance. Specific emphasis must be placed on ensuring the requisite infrastructure, services, and institutions needed for contract enforcement, risk management, grades, standards setting and implementation, and information dissemination. In addition, public policy measures should extend support to producers' organizations to increase producers' bargaining power.

7.7.3 Institutional Innovations

Most producers in developing countries are smallholders who produce a small marketable surplus. Since agricultural value chains are long and dominated by intermediaries, higher transaction and marketing costs hinder the participation of smallholders in markets. The rapid expansion of multinational food companies and supermarkets will likely have further implications for smallholders in developing countries. Being resource-constrained and unable to meet stringent food quality and safety standards, smallholders face difficulties adjusting to the evolving food systems. Therefore, greater emphasis is laid on ensuring smallholders' participation in agricultural markets by developing their informal and formal market linkages (Bbun and Thornton 2013). Institutional innovations such as cooperatives, farmers' associations, and contract farming can help smallholders cross this barrier (Birthal et al. 2007). Producers can improve their bargaining power and get more returns by pooling their produce. Farmers' markets have the potential to reduce inefficiencies caused due to fragmented markets.

7.7.4 Diversification

A notable driving feature of the modern-day agri-food system is the changing preferences and dietary patterns of consumers. Consumers have a considerable dietary shift from staple food to nonstaple, fresh, and processed high-value food items (Gehlhar and Regmi 2005; Grunow and van der Vorst 2010;Reardon and Barrett 2000; Scrinis 2007). As a result, production system diversification from traditional to high-value agricultural produce is rapidly taking place in developed countries (McCullough et al. 2008). The production and exports of high-value agricultural products in fresh and processed forms, fresh produce, seafood, dairy, and poultry products have significantly increased in the last two decades (Athukorala and Jayasuriya 2003; Rae and Josling 2003). Traditional staple food items like cereals are gradually lagging the high-value items in production and per capita availability.

However, producers in developing countries are embracing this transformation at a relatively slower pace. They often grow the same crops for generations and find it difficult to switch over to high-value crops which require learning new production technologies, resources, equipment, and new markets. Therefore, it is necessary to sensitize and facilitate producers in this transformation by providing them with needed awareness, knowledge, resources, and technology.

7.7.5 Standards and Certification Culture

Across the globe, the prevalence of stringent quality grades and standards is gradually growing mainly due to consumer pressure and international regulatory requirements (Chen and Stamoulis 2008). In addition, consumer awareness about product quality and safety issues has increased, and they demand products that meet specifications (Reardon et al. 2001; Trienekens and Zuurbier 2008). Furthermore, the countries are obliged to comply with WTO agreements like the Agreement on Technical Barriers to Trade (TBT) and Sanitary and Phyto-sanitary Measures (SPS).

Given this change, agribusiness firms worldwide are striving hard to achieve various public and private standard certifications such as ISO 9000 and ISO 14000, Hazard Analysis and Critical Control Points (HACCP), Good Agricultural Practices (GAP), traceability, and Maximum Residue Limits (MRL) testing (Kula et al. 2006). Consumers consider these certifications as a guarantee and consider these certifications while purchasing the products (Blackman and Rivera 2011; Peri 2006; Sahota et al. 2009). Producers and other value chain actors operating in markets can no longer avoid this change. They will have to gradually adopt this standardization and certification culture and upgrade their practices to address consumer requirements adequately.

7.7.6 Use of ICT

Information and communication technologies (ICT) are rapidly penetrating developing countries. Their use in rural areas is also gradually increasing. Emerging digital technologies can greatly contribute to upgrading the structure and functioning of agricultural markets in developing countries. Therefore, there is a need to promote its use in agriculture, particularly in marketing agricultural produce. Developing various applications for guiding producers on the efficient and effective use of inputs, modern technologies, and production and marketing practices can significantly improve the agricultural sector. Producers can sell through these applications to consumers by displaying their products.

Modern communication technologies such as smartphones, online platforms, and mobile applications have increased the access of producers to market information services offered by various domestic and international public and private sector institutions. They can use these technologies to establish a network of connections with market intermediaries and experts, get important information more quickly, and make better decisions about logistics and transportation, pricing and location, supply and demand, product diversification, and input access.

7.8 Conclusion

Markets play an important role in linking buyers and sellers, and they substantially contribute to food security in developing countries. Through generating market signals, markets help producers in deciding what, when, and how to produce and know the quality requirements of consumers. Markets are equally important for business buyers and consumers in finding their desired products for business purposes or ultimate consumption. Unlike developed countries, markets in developing countries are not well organized and lack modern facilities, and their performance is suboptimal. Often, surpluses and shortages create food security problems due to market inefficiencies and distortions. Since food security is a pressing challenge, particularly for developing countries, there is a dire need to improve the availability of well-functioning efficient markets to all stakeholders including producers, traders, and consumers. Entrepreneurial mindset, market governance, institutional innovations, diversification, standards and certification culture, and information communication technologies can significantly contribute to improving the performance of markets.

References

Alabi, M.O. and Ngwenyama, O. (2023). Food security and disruptions of the global food supply chains during COVID-19: building smarter food supply chains for post COVID-19 era. *British Food Journal* 125 (1): 167–185.

Athukorala, P.-C. and Jayasuriya, S. (2003). Food safety issues, trade and WTO rules: a developing country perspective. *World Economy* 26 (9): 1395–1416.

Barreti, C.B. and Mutambatsere, E. (2008). Agricultural markets in developing countries. In: *The New Palgrave Dictionary of Economics: Volume 1 – 8* (ed. S.N. Durlauf and L.E. Blume), 91–95. London: Palgrave Macmillan UK.

Bbun, T.M. and Thornton, A. (2013). A level playing field? Improving market availability and access for small scale producers in Johannesburg, South Africa. *Applied Geography* 36: 40–48.

Birthal, P.S., Jha, A.K., and Singh, H. (2007). Linking farmers to markets for high-value agricultural commodities. *Agricultural Economics Research Review* 20: 425–439.

Blackman, A. and Rivera, J. (2011). Producer-level benefits of sustainability certification. *Conservation Biology* 25 (6): 1176–1185.

Chamhuri, N. and Batt, P.J. (2013). Exploring the factors influencing consumers' choice of retail store when purchasing fresh meat in Malaysia. *International Food and Agribusiness Management Review* 16 (3): 99–122.

Chávez-Dulanto, P.N., Thiry, A.A.A., Glorio-Paulet, P. et al. (2021). Increasing the impact of science and technology to provide more people with healthier and safer food. *Food and Energy Security* 10 (1): e259.

Chen, K.Z. and Stamoulis, K.G. (2008). The changing nature and structure of agri-food systems in developing countries: beyond the farm gate. In: *The Transformation of Agri-food System: Globalisation, Supply Chains and Smallholder Farmers* (ed. E.B. McCullough, P.L. Pingali, and K.G. Stamoulis), 143–157. London, UK: The Food and Agriculture Organisation of the United Nations (FAO) and Earthscan.

Gehlhar, M. and Regmi, A. (2005). Factors shaping global food markets. In: *New directions in global food markets, Agriculture Information Bulletin No.794* (ed. A. Regmi and M. Gehlhar), 5–17. United States Department of Agriculture (USDA).

Ghani, J.A. (2005). Consolidation in Pakistan's retail sector. *Asian Journal of Management Cases* 2 (2): 137–161.

Grunow, M. and van der Vorst, J. (2010). Food production and supply chain management. *OR Spectrum* 32 (4): 861–862.

Hussain, I. (2018). *Governing the Ungovernable: Institutional Reforms for Democratic Governance*. Pakistan: Oxford University Press.

IFPRI (2018). *2018 Global Food Policy Report*. Washington, DC: International Food Policy Research Institute.

Kula, O., Downing, J., and Field, M. (2006). *Globalization and the Small Firm: AValue Chain Approach to Economic Growth and Poverty Reduction*, 42. Washington, DC, USA: United States Agency for International Development (USAID).

Lencucha, R., Pal, N.E., Appau, A. et al. (2020). Government policy and agricultural production: a scoping review to inform research and policy on healthy agricultural commodities. *Globalization and Health* 16 (1): 11.

Macharia, J., Collins, R., and Sun, T. (2013). Value-based consumer segmentation: the key to sustainable agri-food chains. *British Food Journal* 115 (9): 1313–1328.

McCullough, E.B., Pingali, P.L., and Stamoulis, K.G. (2008). Small farms and the transformation of food systems: an overview. In: *The Transformation of Agri-food Systems: Globalisation, Supply Chains and Farmers* (ed. E.B. McCullough, P.L. Pingali, and K.G. Stamoulis), 3–46. London, UK: Food and Agriculture Organisation (FAO) of the United Nations and Earthscan.

Mohy-ud-Din, Q. and Badar, H. (2011). *Marketing of Agricultural Products in Pakistan: Theory & Practice*. Islamabad: Higher Education Commission of Pakistan.

Njuki, J., Kaaria, S., Chamunorwa, A., and Chiuri, W. (2011). Linking smallholder farmers to markets, gender and intra-household dynamics: does the choice of commodity matter? *The European Journal of Development Research* 23 (3): 426–443.

Peri, C. (2006). The universe of food quality. *Food Quality and Preference* 17 (1–2): 3–8.

Rae, A. and Josling, T. (2003). Processed food trade and developing countries: protection and trade liberalization. *Food Policy* 28 (2): 147–166.

Randell, H., Gray, C., and Shayo, E.H. (2022). Climatic conditions and household food security: evidence from Tanzania. *Food Policy* 112: 102362.

Reardon, T. and Barrett, C.B. (2000). Agroindustrialization, globalization, and international development: an overview of issues, patterns, and determinants. *Agricultural Economics* 23 (3): 195–205.

Reardon, T., Codron, J.-M., Buschc, L. et al. (2001). Global change in agri-food grade and standards: agribusiness strategic responses in developing countries. *International Food and Agribusiness Management Review* 2 (3/4): 421–435.

Sahota, A., Haumann, B., Givens, H., and Baldwin, C. (2009). Ecolebeling and consumer interest in sustainable products. In: *Sustainability in the Food Industry* (ed. C. Baldwin). Wiley-Blackwell and the Institute of Food Technologists.

Scrinis, G. (2007). From techno-corporate food to alternative agri-food movements. *Local-Global* 4: 112–140.

Senauer, B. and Venturini, L. (2005). The globalisation of food systems: a conceptual framework and empirical patterns. Working paper no. 05-01, The Food Industry Center, University of Minnesota.

Timmer, C.P. (2014). The dynamics of agricultural development and food security in Southeast Asia. In: *Routledge Handbook of Southeast Asian Economics*. Routledge.

Traill, B. (1997). Globalisation in the food industries? *European Review of Agricultural Economics* 24: 390–410.

Trienekens, J. and Zuurbier, P. (2008). Quality and safety standards in the food industry, developments and challenges. *International Journal of Production Economics* 113 (1): 107–122.

Van der Meer, C.L.J. (2006). Exclusion of small farmers from coordinated supply chains: market failures, policy failures or just economies of scale? In: *Agri-Food Supply Chains and Networks for Development*, Wageningen UR Frontis Series, vol. 14. Dordrecht: Springer.

Wang, X. (2022). Managing land carrying capacity: key to achieving sustainable production systems for food security. *Land* 11 (4): 484–501.

8

Price Volatility and Food Security: Measurement and Implications

Burhan Ahmad[1], Hammad Badar[1], Ole Gjølberg[2], and Mubashir Mehdi[3]

[1] *Institute of Business Management Sciences, University of Agriculture, Faisalabad, Pakistan*
[2] *NMBU School of Economics and Business, Norwegian University of Life Sciences (NMBU), Aas, Norway*
[3] *Department of Agri-Business and Applied Economics, MNS University of Agriculture, Multan, Pakistan*

8.1 Introduction

Prices possess an important place in efficient allocation of resources by both producers and consumers. Production and consumption decisions by producers and consumers, respectively, are based on the price's information. Prices directly affect the producer's profit, incentive to produce, and consumer's purchasing power (economic access). Price volatility, variation in the prices, also affects the decisions of producers and consumers (Díaz-Bonilla 2016). Price volatility illustrates the degree of price variation or the risk of large, unexpected changes in prices. The risk of severe price events can lead to higher social risks with respect to food security (Kalkuhl et al. 2016). In a broad sense, price volatility refers to movements around a more or less stable price trend or long-term price (Hull 2012). These price variations which are short-term in nature may refer to prices measured on daily, weekly, or monthly basis. Time periods when commodity prices reach to very high or very low level reflect the periods of crises as they create a challenge for producers and consumers as well as for policymakers to make efficient and effective decisions (Kalkuhl et al. 2016).

Food Security in the Developing World, First Edition. Edited by Muhammad Khalid Bashir, Steven G.M. Schilizzi, and Ghaffar Ali.

Commodity prices tend to be volatile, and volatility in the case of agricultural commodities is higher than the other commodities because of its differentiating features, such as being dependent on environment (Deaton and Laroque 1992, Newbery 1989, Pindyck 2004). High food price volatility creates difficulties in making the prediction of prices changes of agricultural commodities which might wield large effects particularly on developing economies which depend on the production, export, and import of food commodities. Prolonged episodes of high food price volatility may lead to increase in economic inequality and fortifying poverty traps (Zimmerman and Carter 2003 in Rapsomanikis 2011). The protracted periods of high volatility in the prices of agricultural commodities raise concerns for the producers, consumers, traders, and governments (Kroner et al. 1995).

Food price volatility is one of the key concerns for policymakers and development practitioners all over the world, particularly in developing countries. For the last many years, food prices in the international as well as domestic markets spiked multiple times and showed large fluctuations leading to high price volatility. This high food price volatility has important implications particularly for the poor people spending larger proportion of their income on food items (Banerjee and Duflo 2008). Moreover, food price volatility can jeopardize the macroeconomic stability and can lead to reduction in incentives of producers for food items (Haile et al. 2014). However, production of food items needs to be increased to fulfill a growing demand for grains and ensure food security because of increase in population (Kornher and Kalkuhl 2013). Hence, food price volatility poses problems for making the effective macroeconomic as well as the microeconomic policy (Deaton and Laroque 1992, Stigler 2011).

Several policy debates regarding adequate prices of food commodities and controlling for price volatility have been in place. These policy debates focused on addressing the interests of both producers (profit maximizations) and consumers (food security). In the case of agricultural and food policies, there have been several debates about adequate price levels of food products and ways of reducing price volatility to a degree that does not interfere with the signaling effects of prices for economic decisions. Those policy issues revolve around balancing the interests of producers and consumers in both industrialized and developing countries (Díaz-Bonilla 2016, Timmer 1989). The main reason behind such debate is the varying nature of policy discussion moving around the traditional policy dilemma, that is, high prices to promote higher production and low prices to increase consumption. Higher prices of agricultural and food prices provide incentives to producers to produce more ensuring future availability; however, this may lead to reduced economic access of consumers to food. On the other hand, low prices can increase food security but may lead to low future supply (Díaz-Bonilla 2016, Timmer et al. 1983, Timmer 1989).

Price levels remained the major focus of much of the policy debate in the world during the 1990s, while price spikes along with ups and downs in the prices from 2008 to 2011 led to shift in policy discussion to price volatility and effects of extremely high prices. In particular, sharp increase in the prices of food commodities in 2007/2008 and as a result the trade policies adopted by various countries, along with the stagnation of the Doha Round negotiations, have invigorated interest in analyzing various trade issues and food security, particularly in developing countries. Moreover, the global financial crisis along with these periods of high price and volatility particularly of some cereals (e.g. rice, the main staple of most of the countries particularly in Asia) have reinforced those concerns (Díaz-Bonilla and Ron 2010a, 2010b, Gilbert and Morgan 2010).

High food price volatility became a burning issue from late 2006 to mid-2008 during food crises 2007–2008 and also on account of high prices in 2011, leading to attract more attention of researchers and policymakers. World dollar prices of major agricultural food commodities rose dramatically from late 2006 through to mid-2008. Sharp decline in prices can be viewed in the second half of 2008 with the beginning of the financial crisis. This occurrence is often termed as the "2008 price spike" (Gilbert and Morgan 2010). The World Bank (2009) stated that "high volatility in the food prices combined with the impact of financial crisis, threatens to further increase food insecurity." In times of crisis, volatility may be self-leading, i.e. generating surge of volatility. Such a phenomenon can lead to "herd-like" behavior where market agents make decisions, following the price trends instead of sticking to market fundamentals (Rapsomanikis 2011). Hence, a better understanding of the food price volatility is a prerequisite for developing strategies to reduce the negative effects from high price volatility and also to devise policies aiming at stabilizing food prices and reducing food insecurity.

Similar kind of questions have been under discussion during the periods of high food price volatility in the past; however, these high volatility periods have remained relatively short, while episodes of market tranquility stayed for a longer time. Therefore, it would not be appropriate to translate current or recent price shocks and high volatility into future; however, it is important to know whether the increase in volatility is permanent (Gilbert and Morgan 2010). Hence, it is important to understand the nature of prices and their volatility for policy analysis by answering relevant question (Díaz-Bonilla and Ron 2010a), such as what is to be measured considering appropriate time frame and properties of the time series data and how to measure, i.e. which one is the appropriate method for measurement taking into account the trends in data series and detrending methods (see, for instance, Canova 1998, 1999).

The rest of the chapter focuses on implications of price volatility for various stakeholders, especially for food security and measurement of price volatility. The case study of rice price volatility in Pakistan is presented as an empirical application. Various policy measures to control the price volatility are discussed at the end.

8.2 Implications of Food Price Volatility

Price volatility raises concerns for various stakeholders and affects their decision-making. It also affects the market functioning and food security. This section presents such implications.

8.2.1 Price Volatility and Decision-Making by Stakeholders

Price volatility exerts effects on the behavior of producers, market intermediaries such as traders and processors, consumers, and governments and these decisions can affect food security. These stakeholders might face problems in dealing with high volatility which can lead to inefficient allocation of resources. High volatility might convey signals of price risk to the investors which in turn discourages investors to invest leading to low production and ultimately to food insecurity. Prolonged episodes of commodity price volatility can affect the economic growth of the country (Ramey and Ramey 1995, Van der Ploeg and Poelhekke 2009), particularly for those countries not having well-developed financial institutions (Aghion et al. 2009).

For producers/farmers, high levels of price volatility can result in inefficient allocation of resources, particularly the use of inputs. In the presence of high volatility, it is difficult for farmers to take efficient decisions for the optimal use of inputs and most likely they use fewer inputs such as fertilizers and low-quality seeds leading to low production levels. Moreover, it may discourage them to invest to increase the productivity resulting in lower incomes and decrease in overall availability food (Kalkuhl et al. 2016, Taya 2012, Assefa et al. 2017). Moreover, price volatility of agricultural inputs render the downstream sector of food supply chains to source uncertainties, leading food and agricultural companies to modify their sourcing strategies as a coping system (Rabobank 2011). On the consumer end of food supply chains, unanticipated surges in prices create food security risks particularly for the consumers having higher expenditure on food items (Hernandez et al. 2014).

Implications of high price volatility for the inventory holders can be in the form of high storage which in turn increases the cost of inventories. Inventories can reduce price volatility, but this is associated with the ability of inventory holders regarding the right prediction of prices in future. On the other hand, if wrong predictions are made then increase in the volume of inventories may lead to higher price volatility (Ahmad et al. 2017).

8.2.2 Price Volatility and Market Integration

Market integration is referred to the degree to which demand and supply situation of one market is transmitted to another market (Negassa and Meyers 2007). Vertical and spatial market integration are two types of market integration (Meyer 2004). Spatial market integration can be attributed to short-term co-movements of prices as well as the existence of long-run relationship among prices of different markets. It can be described as the smooth transmission of price signals and information among geographically separated markets. Vertical market integration can be defined as transmission of price signals from one marketing channel to another (Goletti, Ahmed, & Farid 1995, Mukeere 2009, Minot 2011, Ahmad and Gjølberg 2015, Kabbiri 2016). Agricultural commodity price transmission helps in efficient allocation of resources by the concerned stakeholders and promoting economic growth (Baulch 1997, World Bank 2012). Lack of market integration reflects the existence of infrastructural and institutional bottlenecks. In case food markets are not integrated, price signals are not going to be transmitted from the regions which are deficit in supply to surplus markets and any local food scarcity can persist longer because distant markets (having no scarcity) could not respond to the price signals of food surplus markets (Goletti & Babu 1994, Baulch 1997, Kabbiri 2016). Other benefits of integrated market could be increase in supply and numbers of suppliers, increase in effectiveness of macro-economic policies and reduced price risks (Mukeere 2009)

Increase in market integration can increase price volatility; however, evidence in favor of this notion is limited. The relationship between market integration and price volatility is complex. Market integration can result in higher price volatility as it facilitates the transmission of shocks across different markets. Conversely, market integration can decrease price volatility by offering producers and consumers broader market access and supporting them in doing efficient allocation of resources. While it is logical to assume that markets with high price transmission would also have high volatility transmission, this is not always the case. For example, prices from highly volatile global markets may be transmitted to local markets with a delay of one to six months, shielding local markets from international turmoil and resulting in lower volatility. Additionally, even if there is no direct price transmission, local market volatility can be influenced by the level of uncertainty among local traders, which a sudden increase in global market volatility could impact. Market shocks can cause price and volatility changes, making traditional risk management tools ineffective. Since price volatility can harm consumers and producers, assessing both price and volatility transmission in food markets is very important (Newbery 1989, Serra 2011, von Cramon-Taubadel and Goodwin 2021). Links between geographically separated markets can be formed through trade. Surpluses of commodities such as excess supply of grains are exported, while in time of higher demand than supply (shortage) these are imported helping to stabilize prices and reduce price volatility. However, international markets for wheat, rice, maize, and soybeans, staple foods, are very thin as only a small portion of production of these crops is internationally traded because of high domestic consumption. Top five producers of maize and rice share about 70% of total world production, and top five exporters of these crops captures about 80% share of total global exports of these crops. Similar figures for wheat are 50% (Production) and 60% (Exports). This concentration in few countries indicates a limited world's capacity to comply with the shocks in these countries as any production shock that may be on account of weather shock could have vital effects on prices both in domestic and international markets and on price volatility (Díaz-Bonilla 2016, Jacks et al. 2011).

Market integration may decrease the price volatility, but it contributes to higher spillovers of price volatility across markets which pose problems for policy response to volatility as the nature of the casual effect becomes more complex and affects macroeconomic environment. The use of stocks to fill the gap in supply and demand cannot be focused only rather, the government does efforts to improve the performance of markets by providing information services and infrastructure (Kornher and Kalkuhl 2013). Moreover, monetary policy can affect financial investments, trade, and commodity storage through changes in interest rates, inflation rates, and exchange rates (Frankel 2006). In the same connection, rule of law, contract enforcement, and efficient government administration can facilitate intermediaries to offer insurance and provide capital to encourage efficient allocation of resources and risk assessment (Conning and Udry 2007, Levine et al. 2000). Rules of the governments regarding trade are formulated to facilitate private traders and stockholders to do their businesses and smooth prices through tapping the possibilities of arbitrage.

8.2.3 Price Volatility and Food Security

Food price volatility is closely linked with and exerts effects on food security. Food security can be illustrated with its four main factors which are availability of food, physical and economic access to food, food utilization, and its stability (vulnerability and shocks) over time (FAO 1996, 2015). Generally, food security refers to a state whereby all the people have economic and physical access to a disease-free healthy (safe) and quality food at all the time to fulfill their dietary needs (FAO 1996). The concept of nutrition security is deeper and more requiring than food security as it focuses on access to that nutrition level, which can lead to better performance of the people meaning higher productivity of labor, strengthen the immune system of human body, help to recover from diseases and bearing pregnancy, etc. Low levels of food security (food insecurity) and inadequate nutrition can negatively affect the health and hence human capital, decreasing labor productivity and ultimately leading to low economic output (Behrman and Rosenzweig 2004, FAO 2015, Fogel 1994, Gyimah-Brempong and Wilson 2004, Weil 2007).

Food and nutrition security are among the high-priority areas of the Sustainable Development Goals (SDGs) of the post-2015 development agenda. Various international organizations and governments put lot of efforts and formulated various policies to increase the level of food and nutrition security but in spite of all these, the number of nutrition and food-insecure people remained high. About 2 billion people do not have adequate access to food, while 780 million people are undernourished (Concern 2023, FAO 2015).

The slow pace of reducing hunger and malnutrition coupled with price vitality on account of the food crises of 2007/2008 and 2010 strengthen the concerns regarding the stability of the global food system as well as its reliability (Kalkuhl et al. 2016). It is the responsibility of governments, particularly the governments of developing countries entitle to higher responsibility, to ensure food security at least to a minimum extent contributing to reasonable living conditions. Failure of provision of these basic food requirements leads to protest by the people, and these conditions may lead to reduction in the legitimacy of the governments and unrest in the country particularly in the urban areas of the country as it is relatively easy to organize a protest in the urban areas. Such circumstances require serious political and institutional reforms (Costello et al. 2015).

Prices and changes in prices are connected with the changes in food security. It is very important to understand that how prices affect food security. One reason is that prices are associated with many factors leading to food security such as supply of food commodities, real incomes, and linkages among various markets. Another reason is that data on prices are relatively available in greater frequency and at low cost compared with data on indicators of food security. Moreover, prices can be employed to form future expectations regarding changes in market conditions and associated risk by various stakeholders in the supply chain which are helpful to researchers to explore information processing capacity of markets at greater extent (Fama 1970).

Short-term deficiencies in physical and economic access to food can escort to long-term nutritional loss particularly in children, and this loss may not be reversible. For instance, on account of higher prices (and resulting higher volatility) during 2007–2008, calorie intake in many Latin American countries reduced at national as well as at household level particularly in children of less than two years of age and belonging to the poor families (Robles and Torero 2010). In all of the Latin American countries, poor household and already vulnerable families were exposed to higher nutritional deficiency. Higher level of prices of food commodities in Mozambique raised the number of underweight children (Arndt et al. 2012). Similarly in El Salvador, stunting of childhood was increased because of higher food prices in 2007–2008 (de Brauw 2011). Child growth was affected by poor harvests and adverse weather conditions in Zimbabwe (Hoddinott and Kinsey 2001) and leaded to decrease in children weight in Côte d'Ivoire (Jensen 2000). This also reduced the blood concentration of vitamin A and vitamin E in mothers in Zambia (Gitau et al. 2005). Nutritional deficiency can exert long-term effects on the health of the people and can lead to cognitive capabilities (Victora et al. 2008). Malnutrition such as lack of intake of vitamins (vitamin A) can lead to higher probability of permanents disabilities (Black et al. 2008) hence effective food policy is very important.

Food policy is considered as a sensitive and very important political issue, and its sensitivity is increasing with the increasing urbanization because urbanization brings the voice of people closer to the people having power of policy-making. Thinking about the goals such as adequate nutrition and food security as well as ways to attain these goals affects the formulation of food policy. Appropriate suggested policy measures depend on the specific context under consideration (Pinstrup-Andersen 2015). Rising food prices as well as price volatility provoke the protests against governments which affects the political powers of the leaders. For instance, increase in the prices of onions completely changed the picture of outcomes on elections in India. Thousands of protesting people came into the streets of Port au Prince (in 2008), in Algiers (in 2011), and in several Arab countries (in 2008). Haitian prime minister had to resign on account of these high food prices. Similar kind of examples are found in other countries leading to social unrest, and findings of some empirical studies show a causality between food prices and social unrest (Bellemare 2015).

Price volatility may lead to food insecurity and inadequate micro- and macro-nutrient consumption which in turn can reduce physical and mental growth of the children and low labor productivity, hence food policy to control price volatility and food insecurity are essential (Black et al. 2013). Many governments particularly of the developing countries need to focus on formulating effective policies to reduce the price risk and food and nutrient insecurity. However, important policy measures which can trim down the price volatility and decrease the food insecurity may include opening up the international trade through liberalized trade policies, flexible policies for bio-energy, building grain reserves and regulating the commodity markets. Policymakers can espouse a portfolio of policies instead of choosing one policy or the other. This selection is expected to depend on context-specific and country-specific issues under consideration (Kalkuhl et al. 2016).

8.3 Measuring Price Volatility

Various methods are used to measure price volatility. Generally, price volatility measures can be categorized into two perspectives. One perspective is historical, i.e. ex post and another is forward-looking, i.e. ex ante. Ex post measurement of volatility is based on the realized volatility and is termed as measure of unconditional volatility which does not take into account the lagged volatility and lagged prices. Whereas, ex ante measures of volatility employ lagged prices and lagged volatilities and referred to conditional volatility measures. These measures employ information available at time t and predict price volatility at time $t+1$. Conditional volatility measures keep on changing over time; therefore, these are dynamic in nature and forward-looking enabling them to represent varying degree of risk perception (Huchet-Bourdon 2011).

Choosing an appropriate measure of volatility (conditional or unconditional) depends on the specific context of the underlying research, availability of the data, and specific research objectives/questions. Unconditional or ex post measures of volatility can be applied on time series data even with low number of observations. These measures can also be applied on the data with missing observations which is an issue in the time series data availability particularly in developing countries. Unconditional measures of calculating price volatility serve as appropriate tools to examine the effects of realized past shocks. Economist usually employ standard deviations of changes in prices in the logarithmic form usually called price returns to calculate unconditional price volatility, since it is a unit-free measure (Gilbert and Morgan 2010, Kellard and Wohar 2006, Kim et al. 2003).

8.3.1 Unconditional Price Volatility

As unconditional volatility measures assume of a constant variance, they do not explicitly model the historical development of volatility or future expectations regarding price risks. Unconditional measures of volatility are therefore of limited use for the purpose of predicting price volatility. These are also of limited use for modeling risk perceptions of forward-looking agents.

Coefficient of variation (CV) is used to measure variation. CV is defined as the ratio of standard deviation (s) of a variable, price in the case of price volatility, and its mean (I) value over the concerned period, $\mathrm{CV} = s/I$. Estimation of CV for measuring the price volatility has a disadvantage. This indicates that in practical terms, these measures of variability depend on to total time period covered in the data of price series.

Unconditional price volatility can also be measured through calculating standard deviation of price returns. Price returns are usually defined as the deviation between logarithmic prices of one time period and that of next time period. Mathematically, defining price returns r as equal to $(\ln P_2 - \ln P_1)$, it can be written as $\mathrm{SD}(r) = \mathrm{SD}(\ln P_2 - \ln P_1)$. If prices are nonstationary, exhibiting unit root or random walk behavior, with a multiplicative error term, then price returns will be stationary. Hence, standard deviation of these price returns will not be dependent on the sample size. This concept is unconditional in that it does not consider any prior information and is based only on observed variation in price returns.

8.3.2 Conditional Price Volatility

In contrast to ex post measures, conditional price volatility measures typically entail more data and model time series data, which may limit their applicability when data is sparse. However, the choice of the appropriate volatility measure also depends on how agents form their expectations about future price risk. Generally Generalized Autoregressive Conditional Heteroskedastic (GARCH) methods are widely used as forward-looking conditional volatility measures. These models take into account previous years volatility and price shocks (Hernandez et al. 2013, Hull 2012, Rapsomanikis and Mugera 2011). Engle (1982) developed Autoregressive

Conditional Heteroskedasticity (ARCH) models, while Bollerslev (1986) propounded GARCH. A GARCH (p,q) model can be written as follows:

$$\sigma_t^2 = \alpha_0 + \sum_{l}^{m} \alpha_l \mu_{t-l}^2 + \sum_{k}^{s} \beta_k \sigma_{t-j}^2 \tag{8.1}$$

A typical GARCH (1,1) can be written as follows, where $p = q = 1$:

$$\sigma_{t|}^2 = \alpha_0 + \alpha_1 \mu_{t-1}^2 + \beta_1 \sigma_{t-1}^2 \tag{8.2}$$

where σ^2 is the conditional variance from the conditional mean equation, μ_t^2 is the squared error term from the equation for conditional mean, i denotes markets, and t denotes time periods. The advantage of GARCH model is that it allows the variance of returns to vary over time as a function of lagged variance and lagged squared residuals (Ahmad et al. 2017, Bloznelis 2016, Minot 2014). The sum of alphas and betas is represented by the lambda (λ) and for GARCH (m,s) it can be written as $\lambda = \sum_l^m \alpha_l + \sum_k^s \beta_k$ and for GARCH(1,1) it can be written as $\lambda = \alpha_1 + \beta_1$. Lambda λ depicts the persistence parameter and is used as a measure of the persistence in volatility. A value of λ close to 1 shows that the conditional variance will take longer time for reverting to its long-term value of the mean. If $\lambda = 1$, then the conditional variance can be increased with no limits and no inclination for reverting to its long-term mean. The greater than one value lambda ($\lambda > 1$) reveals an explosive growth[1] (Ahmad et al. 2017, Bloznelis 2016).

Multivariate GARCH models are also used to measure the conditional volatility particularity to allow study the spillovers of volatility between different markets and/or across commodities (Hernandez et al. 2013, Rapsomanikis and Mugera 2011). Multivariate GARCH models are used in studies analyzing the volatilities in time series data of prices and their spillover across markets. The Dynamic Conditional Correlation Generalized Autoregressive Conditionally Heteroskedastic (DCC-GARCH) approach of multivariate GARCH models developed by Engle (2002), in particular, is designed to analyze the dynamics of volatility of a time series under analysis and measures conditional correlations among the various time series under study and transmissions among them. In DCC-GARCH model, the conditional variances are modeled as univariate GARCH models (as described in Equation 8.1 and 8.2) and the conditional covariances are modeled as nonlinear functions of the conditional variances. The

1 More details can be found in Ahmad et al. (2017) and Bloznelis (2016).

conditional quasi-correlation parameters that weight the nonlinear combinations of the conditional variances follow the GARCH-like process specified in Engle (2002). The (DCC) GARCH model is about as flexible as the closely related varying conditional correlation (VCC) GARCH model (Dynamic model), rather more flexible than the constant conditional correlation (CCC) GARCH model (static model) and similarly more parsimonious than the diagonal (VECH) GARCH (static) model (Ahmad et al. 2017, Engle 2002,Gotz et al. 2013).

8.3.3 Price Volatility: A Case of Rice Sector in Pakistan

This case study is taken from Ahmad et al. (2017) and presents the analysis of rice price volatility in major domestic markets of Pakistan. It measures both unconditional and conditional price volatility focusing on identifying the differences in the volatilities across markets. Data on monthly wholesale prices during 1994–2011 of six major wholesale markets of IRRI rice in Pakistan is used for the analysis of rice price volatility. Selected six major markets representing four provinces are Rawalpindi and Multan from Punjab, Sukkur and Hyderabad from Sindh, and Peshawar from Khyber Pakhtunkhwa (KPK) and Quetta from Baluchistan. Price of Thai 5% broken rice is used as international market price. Thai 5% broken rice is considered as a close substitute of IRRI rice in Pakistan (Ahmad and Gjølberg 2015, Ahmad et al. 2017) as it is a coarse grain variety similar to IRRI rice variety of Pakistan.

Rice possesses a keen importance in the agricultural sector as well as in the economy of Pakistan. It is the second largest export commodity (cotton and cotton products bearing first position) and second staple food (wheat being the first staple food) in Pakistan, and therefore it has important implications for both producers as profit maximizes and consumers with respect to food security. Rice production in Pakistan shares about 20% of the total cropped area allocated to the production of food grain crops. It contributes about 0.7% to GDP, whereas its share in the value-added agriculture is about 3.1%. At the global level, production of paddy rice accounts for about 1.3% of total paddy production in the world, while export of milled rice from Pakistan captures about 11% of the total global exports of rice (GoP 2013 2017, UN FAO 2016).

Basmati and IRRI are the two major rice varieties produced in Pakistan and are different in their characteristics. IRRI rice is a coarse grain variety with less aroma compared with Basmati rice which is characterized as fine rice having long grain size and very nice aroma. Sindh Province is the major home for IRRI rice production, while majority of the area under basmati rice cultivation lie in Punjab province (Abdullah et al. 2015). Overall, about 50–60% rice is consumed domestically and rest is exported contributing to world food security as rice is a staple food of many countries in the world. Domestic consumption of Basmati rice is higher than that of IRRI rice; hence, share of IRRI rice exports in total rice exports is greater than that of Basmati rice. Moreover, production and yield of IRRI rice are higher than that of Basmati rice (Ahmad et al. 2017, Ahmad and Gjølberg 2015).

A number of government policies have been in place to regulate domestic and international markets of rice of Pakistan. Export sector was privatized in 1988–1989, and rice exporters association was formed. On account of price spike in 2007–2008, the international market government announced minimum export price policy for the sake of food security. Export subsidies have been given to the exporters in two years from 2002 to 2004. A tariff policy of 10% import tariff is in place which is reduced from 15%. Support price policy has been adopted and continued until 2001–2002. After that, indicative price have been announced occasionally (Ahmad and Garcia 2012[2], REAP 2010, Salam 2009, WTO 2011).

Marketing system of rice is of traditional in nature and dominated by the market intermediaries such as village dealers, commission agents, wholesalers, and retailers. These intermediaries par actually govern the marketing system and seek their interests. They usually provide loan to the farmers particularly to the small farmers who is bound to sell to them on specified terms and conditions which result in low prices received by the farmers. Low prices can be regarded as one of the major problems in the value chain system of rice. High postharvest losses are also a major problem which can lead to low prices on account of decrease supply because of these losses. Other barriers in the product flow include poor quality of roads and transport system, shortage and low speed of transport vehicles, and presence of markets at distant places from production areas. These factors may fuel the price volatility and lead to spatial difference in price volatility across markets (Abdullah et al. 2015, Rehman et al. 2012, TDAP 2016).

8.3.4 Salient Features of Regional Rice Prices Volatility

To get an idea about price volatility in Pakistan as well as in the international market, percentage changes in the average price of domestic markets rice in Pakistan as well as of the international market are shown in Figure 8.1. High volatility can be viewed from the large fluctuations particularly after the year 2007–2008. To measure the volatility and view the trend

---

2 More detail on government policies for rice sector in Pakistan can be found in Ahmad and Garcia (2012).

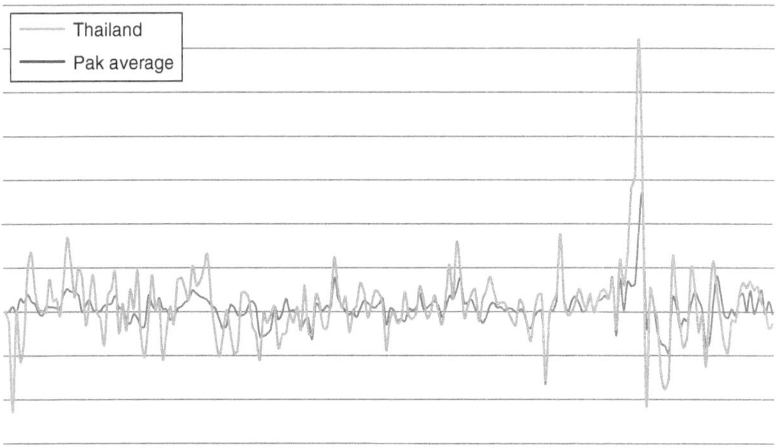

Figure 8.1 Logarithmic price returns in Pakistan's domestic (average) and international rice markets. *Source:* Ahmad et al. (2017).

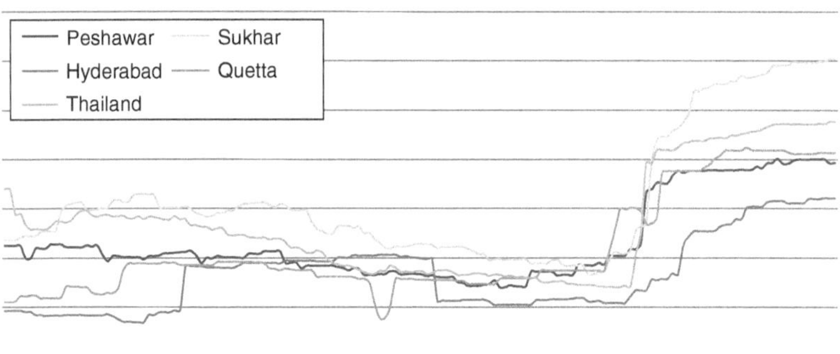

Figure 8.2 Standard deviations of logarithmic price returns in Pakistan's domestic and international rice markets over 48-month rolling windows during 1994–2011. *Note:* Thailand's prices were converted into Pakistan's rupees before estimations of rolling standard deviations.

in volatility, rolling windows of standard deviations of price returns in logarithmic forms for a period of 48 months are estimated and presented in Figure 8.2. As shown in Figure 8.1, high volatility with increasing trend can be viewed from the plotted figure. A sharp increase after the boom-and-bust period 2007–2008 can also be visualized. Higher volatility of rice compared with other food grain crops particular after 2007–2008 was also found by Gilbert and Morgan (2010).

For further examination of trends in rice price volatility, the data under analysis are categorized into three categories based on the three sets of time.

Table 8.1 Standard deviations of logarithmic price changes.

Years	Peshawar	Rawalpindi	Multan	Sukkur	Hyderabad	Quetta
1994–1999	0.044	0.034	0.043	0.056	0.017	0.033
2000–2005	0.033	0.039	0.049	0.043	0.035	0.028
2006–2011	0.072	0.081	0.092	0.104	0.057	0.073
1994–2011	0.052	0.051	0.064	0.071	0.039	0.048

Source: Ahmad et al. (2017)

First subset of the data is for the period 1994–1999, while second category of time periods extends from 2000 to 2005 and third and last category contains price data from 2006 to 2011. Standard deviations of logarithmic rice price returns for the said three categories are estimated, and results are depicted in Table 8.1. These results clearly indicate a rising trend in the rice price volatility providing a support to earlier results. Highest level of rice price volatility is found in the third category of time period from 2006 to 2011. Volatility in this third time increase by two times (doubled) in all the selected markets and even in some markets it increased by more than double amount.

Higher levels of rice price volatility can lead to inefficient allocation of resource by the producers, and inappropriate allocation of budget by the consumers may lead to low profit to farmers and food insecurity for the consumers. Implications of high price volatility for the inventory holders can be in the form of high storage which in turn increases the cost of inventories. Inventories can reduce price volatility, but this is associated with the ability of inventory holders regarding right prediction of prices in future. On the other hand, if wrong predictions are made then increase in the volume of inventories may lead to higher price volatility.

To model the conditional volatility, ARCH/GARCH models are estimated. First of all, to identify the presence of ARCH effects, ARCH-LM tests are applied on logarithmic rice price returns in all the selected markets. The results of the tests indicate that ARCH effects are present in all of the selected domestic markets as well as in the international market. Moreover, spatial differences in volatility across markets can be viewed from the fact that magnitude and significance of the ARCH/GARCH coefficients differ across markets. Markets of Hyderabad and Sukkur show highest volatility persistence. The DCC-GARCH[3] class of models of multivariate GARCH

3 Detailed methodology can be found in Ahmad et al. (2017).

models propounded by Engle (2002) is also employed to study the spatial differences in rice price volatilities across the six selected markets. An important feature of multivariate DCC GARCH model is that it is a dynamic model and captures time-varying effects of volatility and measure transmission of volatility across markets through measuring the conditional correlations (Bloznelis 2016, Got et al.2013). Results of the DCC-GARCH model, presented in Tables 8.2–8.4, show that conditional correlations are statistically significant and positive, indicating positive association of volatility among the sample markets of the study. It means that increase in volatility in one market can lead to increase in volatility in the other market. The results of the conditional correlations also reveal that spatial differences exist among the markets as there is an inverse relationship between conditional correlations and distance between markets. Spatial differences in rice price volatility across markets reveal the existence of differences in infrastructure, communication, and transportation services. It may reflect the market power of the market intermediaries in the value chain system.

Price volatility can be reduced through improving the mechanism of market functioning. These improvements can be made through the formulations of appropriate policies encouraging investments in the physical infrastructure such as making quality roads, effective railway system, and telecommunications. Moreover, negative effects of market power of the market intermediaries can be reduced through increasing the access of both producers and consumers to current and updated market information. This may lead to increase the overall profitability along the entire value chain as well as improve the food security (Ahmad et al. 2017).

Table 8.2 ARCH-LM test on price returns in the domestic rice markets in Pakistan.

Year	Thailand	Peshawar	Rawalpindi	Multan	Sukkur	Hyderabad	Quetta
Skewness	1.0	0.9	7.6	0.8	0.6	0.4	0.8
Kurtosis	2.9	2.6	2.6	2.8	2.5	2.2	2.8
ARCH-LM[a]	15.3	5.5	2.9	30.1	26.0	2.6	12.9

[a] All the coefficients are significant at 1% level of significance except for Rawalpindi and Hyderabad which are significant at 10% level of significance.
Source: Ahmad et al. 2017.

Table 8.3 Univariate part of DCC-GARCH model.

DlnP	Thailand	Hyderabad	Sukkur	Multan	Rawalpindi	Peshawar	Quetta
Constant	0.005	0.009[a]	0.01[a]	0.009 [b]	0.01[a]	0.009[b]	0.01[a]
AR(1)	0.33[a]	0.23[a]	0.21[a]	0.38[a]	0.48[a]	0.29[a]	0.23[a]
ARCH (1)	0.15	0.17[c]	0.20[b]	0.32	0.72[c]	0.21	0.97[a]
GARCH(1)	0.43[b]	0.81[a]	0.72[c]	–	–	0.51[a]	–
Lambda	0.58	0.98	0.92			0.72	
K	1.27	34.31	8.31			2.11	
Constant	0.0003	0.00006[c]	0.0004[b]	0.003[a]	0.001[a]	0.0006[a]	0.0006[a]
ARCH(2)	0.28[b]						
[d]Ljung-Box(3)	4.95[c]	1.22	0.96	2.62	2.19	0.06	3.00
ARCH-LM(3)	4.25	0.49	0.77	1.20	0.13	1.12	0.90

[a/b/c] Statistically significant at the 1%, 5%, and 10% levels, respectively.
[d] Ljung-Box (3 lags) and ARCH-LM (3 lags) tests' statistics for standardized model residuals.
Source: Ahmad et al. 2017.

Table 8.4 Time-varying conditional correlations from DCC-GARCH model in domestic rice markets of Pakistan.

Market pairs	Average conditional correlation	Distance (km)
Rawalpindi – Peshawar	0.17	183
Hyderabad –Sukkur	0.09	323
Quetta – Sukkur	0.18	399
Sukkur – Multan	0.29	468
Multan – Rawalpindi	0.28	548
Quetta – Multan	0.20	625
Multan – Peshawar	0.24	689
Quetta– Hyderabad	0.19	721
Hyderabad –Multan	0.14	781
Quetta – Peshawar	0.16	846
Sukkur – Peshawar	0.18	884
Quetta – Rawalpindi	0.11	902
Sukkur – Rawalpindi	0.23	1012
Hyderabad – Peshawar	0.09	1206
Hyderabad – Rawalpindi	0.05	1325

Source: Ahmad et al. 2017.

8.4 Conclusion

Food policy to control price volatility are very important as price volatility may lead to food insecurity and inadequate micro- and macro-nutrient consumption which in turn can reduce physical and mental growth of the children, low labor productivity, and low economic growth of the country (Black et al. 2013). Uncertainty of expected prices in future can lead to reduction in investment for the production of agricultural commodities which in turn lead to reduced log run food supply. The international communities as well as many governments particularly of the developing countries still need to formulate effective policies to mitigate the price risk. However, excessive price volatility can be reduced by opening up the international trade through liberalized trade policies, by formulating flexible policies for bioenergy, building grain reserves, using at appropriate time and regulating the commodity markets. Improvements in the functioning of the markets through the policies encouraging investments in the physical infrastructure as well as provision of current and updated market information to producers and consumers, leading to efficient allocation of resources, can contribute to reduce the price volatility. Policymakers can adopt a portfolio of policies instead of choosing one policy or the other to solve the relevant problems. Countries having strong administrative power and capacity may focus on social protection, while other may adopt storage policies (Ahmad et al. 2017, Gilbert and Morgan 2010, Kalkuhl et al. 2016).

For policymaking, it is not about choosing one of the policy instruments proposed here, but rather a portfolio of policies that best addresses the relevant issues. The weights of such a portfolio will be context-dependent: countries with high administrative capacity, for example, could rely more on social protection, while others may opt for rule-based storage policies. In any case, policies between countries and domains need to be coordinated to produce synergy and avoid any possible offsetting effects (Gilbert and Morgan 2010, Kalkuhl et al. 2016).

Stockholding of food commodities can be used as a policy instrument to control high food price volatility. However, higher amount of stock may lead to higher volatility. The reason behind this is that under the low levels, relatively small shock to production/consumption can exert larger effects on price and effects could be opposite if stock is in high volumes. Moreover, when stocks are accumulated to high levels, then they stay at high levels until and unless increase in consumption becomes higher than production to utilize past surpluses. Stockholding, therefore, can lead to a cyclical pattern both in prices and volatilities even there are independent shock in supply and demand over time. During 2006, level of world grain stocks reduced to very low levels and regarded as one of the reasons behind the increase in price volatility of grain crops (Gilbert and Morgan 2010). Stockholding can decrease volatility if the stocks build up during excess supply and are disposed of when demand is in excess. However, stockholding can be more useful in decreasing the extent of price falls during the events/period of surplus production compared with decreasing the extent of increase in prices during the shortage, since destocking is associated with the presence of available stocks from the past years. Hence, stockholding can decrease price volatility, but it may lead price distribution to a positive skew (Deaton and Laroque 1992, Wright 2011).

References

Abdullah, M., Li, J., Ghazanfar, S. et al. (2015). Domestic rice marketing structure and marketing margins in Pakistan. *Journal of Economics and Sustainable Development* 6 (12): 226–238.

Aghion, P., Bacchetta, P., Ranciere, R., and Rogoff, K. (2009). Exchange rate volatility and productivity growth: the role of financial development. *Journal of Monetary Economics* 56 (4): 494–513.

Ahmad, B. and Garcia, R.J. (2012). Measuring commodity-specific trade determinants and export potential: a gravity model of Pakistan's rice exports. *Journal of International Agricultural Trade and Development* 2 (8): 125–148.

Ahmad, B. and Gjølberg, O. (2015). Are Pakistan's rice markets integrated domestically and with the international markets? *SAGE Open* 5 (3): 2158244015596791.

Ahmad, B., Gjolberg, O., and Mehdi, M. (2017). Spatial differences in rice price volatility: acasestudy of Pakistan 1994–2011. *The Pakistan Development Review* 56 (3): 265–289.

Arndt,C., Hussain,M.A. and Østerdal,L.P. (2012).Effects of food price shocks on child malnutrition:the Mozambican experience 2008/09 (No. 2012/89). WIDER working paper.

Assefa, T.T., Meuwissen, M.P., Gardebroek, C., and Oude Lansink, A.G. (2017). Price and volatility transmission and market power in the German fresh pork supply chain. *Journal of Agricultural Economics* 68 (3): 861–880.

Banerjee, A.V. and Duflo, E. (2008). What is middle class about the middle classes around the world? *Journal of Economic Perspectives* 22 (2): 3–28.

Behrman, J.R. and Rosenzweig, M.R. (2004). Returns to birthweight. *The Review of Economics and Statistics* 86 (2): 586–601.

Bellemare, M.F. (2015). Rising food prices, food price volatility, and social unrest. *The American Journal of Agricultural Economics* 97 (1): 1–21.

Black, R.E., Allen, L.H., Bhutta, Z.A. et al. (2008). Maternal and child undernutrition: global and regional exposures and health consequences. *Lancet* 371 (9608): 243–260.

Black, R.E., Victora, C.G., Walker, S.P. et al. (2013). Maternal and child undernutrition and overweight in low-income and middle income countries. *Lancet* 382 (9890): 427–451.

Bloznelis, D. (2016). Salmon price volatility: a weight-class-specific multivariate approach. *Aquaculture Economics & Management.* 20 (1): 24–53.

Bollerslev, T. (1986). Generalized autoregressive conditional heteroskedasticity. *Journal of Econometrics* 31 (3): 307–327.

de BrauwA (2011) Migration and child development during the food price crisis in El Salvador. *Food Policy* 36:28–40.

Baulch, B. (1997). Testing for food market integration revisited. *The Journal of Development Studies* 33: 512–534.

Canova, F. (1998). Detrending and business cycle facts. *Journal of Monetary Economics* 41 (3): 475–512.

Canova, F. (1999). Does detrending matter for the determination of the reference cycle and the selection of turning points? *The Economic Journal* 109 (452): 126–150.

Concern (2023). World hunger facts: What you need to know in 2023. Concern Worldwide Organization. https://concernusa.org/news/world-hunger-facts/

Conning, J. and Udry, C. (2007). Rural financial markets in developing countries. *Handbook of Agricultural Economics* 3: 2857–2908.

Costello, M., Jenkins, J.C., and Aly, H. (2015). Bread, justice, or opportunity? The determinants of the Arab awakening protests. *World Development* 67: 90–100.

von Cramon-Taubadel, S. and Goodwin, B.K. (2021). Price transmission in agricultural markets. *Annual Review of Resource Economics* 13: 65–84.

Deaton, A. and Laroque, G. (1992). On the behavior of commodity prices. *The Review of Economic Studies* 59 (1): 1–23.

Díaz-Bonilla, E. and Ron, J.F. (2010a). Food security, price volatility, and trade: some reflections for developing countries. *Issue Paper* 8.http://www20.iadb.org/intal/catalogo/PE/2010/07451.pdf.

Díaz-Bonilla, E. and Ron, J.F. (2010b). Food security, price volatility and trade. *ICTSD Issue Paper* 27. https://www.ictsd.org/sites/default/files/research/2011/12/food-security-price-volatility-and-trade.pdf.

Díaz-Bonilla, E. (2016). Volatile volatility: conceptual and measurement issues related to price trends and volatility. In: *Food Price Volatility and Its Implications for Food Security and Policy* (ed. M. Kalkuhl, J. von Braun, and M. Torero), 35–57. Springer International Publishing.

Engle, R.F. (1982). Autoregressive conditional heteroscedasticity with estimates of the variance of United Kingdom inflation. *Econometrica: Journal of the Econometric Society* 50 (4): 987–1007.

Engle, R.F. (2002). Dynamic conditional correlation: asimple class of multivariate generalized autoregressive conditional heteroscedasticity models. *Journal of Business and Economic Statistics* 20 (3): 339–350.

Fama, E.F. (1970). Efficient capital markets: a review of theory and empirical work. *Journal of Finance* 25 (2): 383–417.

FAO (1996). *Rome Declaration on World Food Security and World Food Summit Plan of Action*. Rome: FAO.

FAO (2015). *The State of Food Insecurity in the World*. Rome: Food and Agriculture Organization of the United Nations (FAO). https://www.fao.org/3/i4646e/i4646e.pdf (accessed 15 November 2023).

Fogel, R.W. (1994). Economic growth, population theory, and physiology: the bearing of long-term processes on the making of economic policy. *American Economic Review* 84: 369–395.

Frankel, J.A. (2006). *The Effect of Monetary Policy on Real Commodity Prices* (No. w12713). National Bureau of Economic Research. https://www.nber.org/system/files/working_papers/w12713/w12713.pdf (accessed 15 November 2023).

Gilbert, C.L. and Morgan, C.W. (2010). Food price volatility. *Biological Sciences* 365 (1554): 3023–3034.

Gitau, R., Makasa, M., Kasonka, L. et al. (2005). Maternal micronutrient status and decreased growth of Zambian infants born during and after the maize price increases resulting from the southern African drought of 2001–2002. *PublicHealth Nutrition* 8 (7): 837–843.

Goletti, F., Ahmed, R., and Farid, N. (1995). Structural determinants of market integration: the case of rice markets in Bangladesh. *Development Economics* 33:196–198.

Goletti, F. and Babu, S. (1994). Market liberalization and integration of maize markets in Malawi. *Agricultural Economics* 11: 311–324.

Gotz, L., Goychuk, K., Glauben, T., and Meyers, W. H. (2013).The dynamics of Ukrainian farm wheat prices volatility: evidence from a dynamic conditional correlation GARCH model development. *Proceedings of the NCC-134 Conference on Applied Commodity Prices Analysis, Forecasting, and Market Risk Management*, St. Louis, MO. https://ageconsearch.umn.edu/record/285798 (accessed 15 November 2023).

Government of Pakistan (GoP) (2013). *Agricultural Statistics of Pakistan (2011–12).* Islamabad, Pakistan: Federal Bureau of Statistics of Pakistan.

Government of Pakistan (GoP) (2017). *Economic Survey of Pakistan (2013–14).* Islamabad, Pakistan: Economic Advisor's Wing, Finance Ministry.

Gyimah-Brempong, K. and Wilson, M. (2004). Health human capital and economic growth in SubSaharan African and OECD countries. *Quartely Review of Economics and Finance* 44 (2): 296–320.

Haile, M.G., Kahkuhl, M., and von Braun, J. (2014). Inter- and intra-seasonal crop acreage response to international food prices and implications for volatility. *Agricultural Economics* 45 (2014): 693–710.

Hernandez, M. A., Ibarra, R., Trupkin, D. R. (2014). How far do shocks move across borders? Examining volatility transmission in major agricultural futures markets. *European Review of Agricultural Economics* 41 (2): 301–325.

Hoddinott, J. and Kinsey, B. (2001). Child growth in the time of drought. *Oxford Bulletin of Economics and Statistics* 63: 409–436.

Huchet-Bourdon, M. (2011). *Agricultural Commodity Price Volatility: An Overview.* OECD Food, Agriculture and Fisheries Working Papers No. 52 https://www.oecd-ilibrary.org/docserver/5kg0t00nrthc-en.pdf?expires=1699987107&id=id&accname=guest&checksum=71B6DCAD33B74A2E64FDF94E0D0674F3.

Hull, J.C. (2012). *Options, Futures and Other Derivatives*, 8e. Essex: Pearson Education.

Jacks, D.S., O'Rourke, K.H., and Williamson, J.G. (2011). Commodity price volatility and world market integration since 1700. *Review ofEconomics and Statistics* 93 (3): 800–813.

Jensen, R. (2000). Agricultural volatility and investments in children. *American Economic Review* 90: 399–404.

Kabbiri, R., Dora, M., Elepu, G., and Gellynck, X. (2016). A global perspective of food market integration: a review. *Agrekon* 55 (1–2): 62–80.

Kalkuhl, M., von Braun, J., and Torero, M. (ed.) (2016). *Food Price Volatility and Its Implications for Food Security and Policy*. Springer Open.

Kellard, N. and Wohar, M.E. (2006). On the prevalence of trends in primary commodity prices. *Journal of Development Economics* 79 (1): 146–167.

Kim, M.H., Lee, E.K., Jun, J.H. et al. (2003). Hydrogen production by catalytic decomposition of methane over activated carbons: deactivation study. *Korean Journal of Chemical Engineering* 20 (5): 835–839.

Kornher, L. and Kalkuhl, M. (2013). Food price volatility in developing countries and its determinants. *Quarterly Journal of International Agriculture* 52 (4): 277–308.

Kroner, K.F., Kneafsey, K.P., and Claessens, S. (1995). Forecasting volatility in commodity markets. *Journal of Forecasting* 14 (2): 77–95.

Levine, R., Loayza, N., and Beck, T. (2000). Financial intermediation and growth: causality and causes. *Journal of Monetary Economics* 46 (1): 31–77.

Meyer, J. (2004). Measuring market integration in the presence of transaction costs–a threshold vector error correction approach. *Agricultural Economics* 31: 327–333.

Minot, N. (2011). In: Transmission of world food price changes to markets in Sub-Saharan Africa (IFPRI Discussion Paper No), 01059–Washington, DC: International Food Policy Research Institute.

Minot, N. (2014). Food price volatility in sub-Saharan Africa: Has it really increased? *Food Policy*, 45, 45–56. https://www.sciencedirect.com/science/article/pii/S0306919213001863 (accessed 15 November 2023).

Mukeere, B. (2009). Basics of market analysis for food security. Available at SSRN 1370865. https://documents.wfp.org/stellent/groups/public/documents/manual_guide_proced/wfp204080.pdf (accessed 15 November 2023).

Negassa, A. and Myers, R. J. (2007). Estimating policy effects on spatial market efficiency: an extension to the parity bounds model. *American Journal of Agricultural Economics* 89: 338–352.

Newbery, D.M. (1989). The theory of food price stabilization. *The Economic Journal* 99 (398): 1065–1082.

Pindyck, R.S. (2004). Volatility and commodity price dynamics. *Journal of Futures Markets* 24 (11): 1029–1047.

Pinstrup-Andersen, P. (ed.) (2015). Food price policy in an era of market instability. A politicaleconomy analysis. In: *WIDER Studies in Development Economics*. Oxford: Oxford University Press.

Rabobank (2011). Rethinking the food and agribusiness supply chain; impact of agricultural price volatility on sourcing strategies. http://hugin.info/133178/R/1549493/476482.pdf (accessed 15 November 2023).

Ramey, G. and Ramey, A. (1995). Cross-country evidence on the link between volatility and growth. *American Economic Review* 85 (5): 1138–1151.

Rapsomanikis, G. (2011). Price transmission and volatility spillovers in food markets. In: *Safeguarding: Food Security in Volatile Global Markets* (ed. A. Prakash), 144–163. Rome, Italy: Food and Agricultural Organization of the United Nations.

Rapsomanikis, G. and Mugera, H. (2011). Price transmission and volatility spillovers in food markets of developing countries. In: *Methods to Analyze Agricultural Commodity Price Volatility* (ed. R. M'barek and I. Piot-Lepetit), 165–179. New York: Springer.

Rehman, A., Muhammad, S., Chaudry, K.M. et al. (2012). Identificationand prioritization of problems faced by rice growers: a case study of Tehsil Daska, Sialkot. *Pakistan Journal of Agricultural Sciences (Pakistan)* 49 (3): 385–390.

Rice Exporters Association of Pakistan (REAP) (2010). Introduction. Rice Exporters Association of Pakistan's Web Page. https://reap.com.pk/aboutREAP (accessed 15 November 2023).

Robles, M. and Torero, M. (2010). Understanding the impact of high food prices in Latin America. *Economia* 10 (2): 117–164.

Salam, A. (2009). Distortions in incentives to production of major crops in Pakistan: 1991–2008. *The Journal of International Agricultural Trade and Development* 5 (2): 185–207.

Serra, T. (2011). Food scare crises and price volatility: the case of the BSE in Spain. *Food policy* 36 (2): 179–185.

Stigler, M. (2011). Commodity prices: theoretical and empirical properties. In: *Safeguarding: Food Security in Volatile Global Markets* (ed. A. Prakash), 25–44. Rome, Italy: Food and Agricultural Organization of the United Nations.

Taya, S. (2012). Stochastic model development and price volatility analysis (OECD food, agricultural and fisheries working papers No. 57). OECD Publishing.

Timmer, P. (1989). Food price policy: the rationale for government intervention. *Food Policy* 14 (1): 17–27.

Timmer, C.P., Falcon, W.P., and Pearson, S.R. (1983). *Food Policy Analysis* (Vol. 1983, pp. 1–301). Baltimore: Johns Hopkins University Press.

Trade Development Authority of Pakistan (TDAP), European Union funded Trade Related Technical Assistance (EU TRTA II) (2016). Sectoral Competitiveness and Value Chain Analysis.

United Nations, Food and Agriculture Organization (UN FAO) (2016). On line data base. https://www.fao.org/faostat/en/#data (accessed 15 November 2023).

Van der Ploeg, F. and Poelhekke, S. (2009). Volatility and the natural resource curse. *Oxfordeconomic papers* 61 (4): 727–760.

Victora, C.G., Adair, L., Fall, C. et al. (2008). Maternal and child under nutrition: consequences for adult health and human capital. *The Lancet* 371: 340–357.

Weil, D.N. (2007). Accounting for the effect of health on economic growth. *Quarterly Journal of Economics* 122 (3): 1265–1306.

World Bank (2009). *Global Food Crises at a Glance*. Issue briefs. The World Bank.

World Trade Organisation (WTO) (2011). Pakistan's domestic support notifications. WTO on line data base. www.wto.org (accessed 15 November 2023).

World Bank (2012). Transmission of global food prices to domestic prices in developing countries: why it matters, how it works, and why it should be enhanced (Contribution from the World Bank to the G20 Commodity Markets Sub Working Group). The World Bank. https://documents1.worldbank.org/curated/en/234511468316442852/pdf/712680WP0WB0CP00Box370065B00PUBLIC0.pdf.

Wright, B.D. (2011). The economics of grain price volatility. *Applied Economic Perspectives and Policy* 33 (1): 32–58.

Zimmerman, F.J. and Carter, M.R. (2003). Asset smoothing, consumption smoothing and the reproduction of inequality under risk and subsistence constraints. *Journal of Development Economics* 71 (2): 233–260.

9

Food Security, Food Safety, and Sanitation

Mian Kamran Sharif[1], *Kainat Sarwar*[1], *Namrah Abid*[1], *and Muhammad Abid Bashir*[2]

[1] *National Institute of Food Science and Technology, University of Agriculture, Faisalabad, Pakistan*
[2] *Department of Surgery, Rabigh Faculty of Medicine, King Abdul Aziz University, Jeddah, Saudi Arabia*

9.1 Introduction

Food is one of the basic needs without which living beings cannot survive. Hence, adequate food that ensures health and well-being is a basic right of every human inhabiting this planet irrespective of any differences as recognized under article 25 of "Universal declaration of human rights" (UN 1948). This means that food should always be available, enough to meet body requirements, and safe to consume. Unfortunately, as the population of the world continues to increase, from 7.7 billion (2019) to 9.7 billion (2050) as predicted by United Nations (UN 2019a), the food requirements to feed this number of people is also increasing. It is unfortunate that even this basic right is not available to every human as 690 million people went to their beds without food in their stomach in 2019 (UN 2019b), and this number is likely to get doubled by the end of 2020 as the world continues to face the pandemic of COVID-19 (WFP 2019). The worsening of food security conditions all around the globe may be attributed to restrictions in production, processing, loss of jobs, closure of markets, and distribution channels, etc. (Gebeyehu et al. 2022). Moreover, recent issues like Russian–Ukraine war, Turkey–Syria earthquake, and Pakistan floods have escalated problems in major parts of world, especially among developing countries. In addition to food unavailability, the world faces another problem of food safety which means that out of the remaining humans who have food, many of them are

Food Security in the Developing World, First Edition. Edited by Muhammad Khalid Bashir, Steven G.M. Schilizzi, and Ghaffar Ali.

at risk of eating unsafe food and millions of them get sick (600 million per year) due to unsafe food out of which some of them even die according to World Health Organization. In other words, even the people who have adequate food are deprived of the basic right of having "safe food" to eat that promotes health and does not put a risk to it. Food safety is integral for food security as unsafe food does more harm than good (Fung et al. 2018). Food safety includes all the possible measures, strategies, and resources to ensure that food remains safe during production, processing, and distribution processes (Raspor et al. 2016). Sanitation and clean water are not only necessary to ensure the production and preparation of safe food but are also basic human rights. It is reported that one in three people do not have safe water to drink and two out of five people do not have the basic facility for handwashing. These figures are alarming as lack of safe food, water, and sanitation mean numerous health hazards and risk of foodborne illnesses (UN 2020).

9.2 What Is Food Safety?

According to FAO, food safety means that the food is safe to consume, and it will not cause any harm to human beings when eaten or prepared according to the intended use (FAO/WHO 2009). Nutritional value is also considered in some broader definitions of food safety (Jevsnik et al. 2008). Food hygiene is also an important term that is usually used when we are talking about food safety. FAO defines it as processes and measures that are taken to ensure the safety of food at all steps of the food supply chain (FAO/WHO 2009). Since safe food is a basic human right, it should be a priority of all the stakeholders and governments to ensure food safety. It is very unfortunate that food safety in our world has become more of a luxury than a necessity as people from underdeveloped and developing countries continue to face food shortages on a regular basis.

9.3 Impact of Food Safety on Food Security

Food safety and food security are interrelated to each other, both being important for humans, but they are not similar terms. Food safety deals with all the aspects to ensure food is free of contamination that does not render illness or injury, whereas food security deals with the availability and access of food (National Geographic 2023). Food security aims at the availability, access, and utilization of "safe" food which is incomplete without food safety. Food safety positively impacts food security through decreasing the economic burden on countries or households that can allow them to better access food. Food safety is a major concern for developed countries due to which sometimes even safe food gets wasted. By improving labeling legislations and employing active packaging, we can save up to 20% of the food that is wasted by the consumers in developed countries like United States (FDA 2022). Out of 48 million annual foodborne related illnesses in US, only 20% get diagnosed with a specific pathogen as the causative agent. Only these 20% specifically diagnosed cases cost about 15.5 billion dollars which poses a huge burden economically (Hoffmann et al., 2015). Strict trade regulations across the borders to prevent disease outbreaks are important for ensuring food safety. It is important for the sellers from developing countries to follow these rules, but at the same time lack of resources or lack of affordability hinders the trade and market access of these food items in return. The cost of food safety measures is a major constraint, especially in lower- and middle-income countries, for food security. Preventive approach to food contamination and hygienic measures can be an efficient way of dealing with food safety (Hoffmann et al. 2019). Food security and food safety go hand in hand. Therefore, research based on both aspects of food systems should be done to find affordable, efficient, and sustainable ways of ensuring them followed by the modifications in food policies and systems accordingly. There is need to focus on food safety as being an important pillar of food security, and how it impacts other sustainable development goals by improving water quality, hunger, poverty, and health and aiding in sustainability and economics of the world.

9.4 Major Problems and Challenges in Food Safety

With recent developments in agriculture, technology, and food supply chain, the ways to ensure food safety have also changed. The innovation in agricultural practices has decreased the time taken for food to reach from farm to plate. According to Avgoustaki and Xydis (2020), modern farming techniques like indoor vertical farming and greenhouses can protect crops better from viruses, pathogens, and harmful chemicals due to controlled environment. Digitalization of agriculture has reduced the use of fertilizers and insecticides (Talaviya et al. 2020). Food technologists are working on antimicrobial and intelligent packages free of harmful chemicals. Digital innovations have also directly facilitated the methods of food safety assurance and surveillance through ensuring real-time monitoring and tracking of food (Xi et al. 2021). Although modern technology has improved food safety measures, the new era is posing new challenges in the way of food safety at the same time. Below are some challenges that are being faced by governments and people in the food industry today.

9.4.1 Food Hygiene

Food hygiene may be compromised due to several reasons such as contamination of food, inappropriate food processing, storing, or cooking techniques, temperature, and pH. If food hygiene is compromised for any of the reasons, food becomes unsafe and poses health risks (Ucar et al. 2016).

Contamination of food: Food contamination is the basic problem that affects food hygiene. Contamination can occur at any level from production to processing or even at consumption stage. **A contaminant** is basically any foreign substance that gets added into food unintentionally and compromises its safety and quality as a result (FAO/WHO 2009).

There are main types of contaminants:

1) **Biological contaminants:** Living contaminants that get added to food unintentionally are called biological contaminants. These include organisms like bacteria, parasites, and viruses that pose serious health risks to human beings. Bacterial contamination is found to be the most common cause of food poisoning around the globe.
2) **Physical contaminants:** Any foreign object that contaminates food at any stage is called a physical contaminant. These objects can injure someone or may act as a source of biological contaminants. Band-Aids, pieces of equipment, nails, and hair are common examples of physical contaminants (Australian Institute of Food Safety 2020).
3) **Chemical contaminants:** Chemical contaminants as the name indicate the unintentional presence of chemicals that either should not be present in food or their concentration in food is above prescribed level. Chemical contamination can occur from products like plastics, detergents, disinfectants, and pesticides. Even water may sometimes act as a source of chemical contaminants. Overuse of some food additives, colors, and preservatives may also lead to chemical contamination (Rather et al. 2017). Ingestion of food with chemical contaminants like heavy metals and pesticides can cause gastrointestinal infections (Song et al. 2017).

9.4.2 Impact of Climate Change

Climate change is the major issue the world is facing today, and researchers think that this will affect food security and food safety too. Due to increasing human population and because of this increasing human activities, greenhouse gases are increasing causing an increase in temperature around the globe. An increase in temperature around the world means an increase in the speed of the water vapor cycle. This will result in floods in some areas and desertification in others (Manabe 2019). Changes in temperature, weather patterns, and precipitation have direct impact on agriculture practices, thus raising concerns for food safety (Herrera et al. 2016).

High temperature affects microbial and viral occurrence and growth patterns. It also affects the occurrence of their corresponding foodborne diseases. Most foodborne bacteria can grow at room temperature. An increase in temperature above that is found to elevate their growth. So global warming around the world may provide a better chance for bacteria causing foodborne diseases to proliferate, thus posing a risk to food safety (Tirado et al. 2010). Growth of pathogen *Trypanosoma cruzi* is also found to be increased by elevating temperatures (Asin and Catala 1995). It was found that episodes of diarrhea increased by 8% for every degree temperature increase in Peru (Checkley et al. 2000). Some climate change studies in Australia also predicted an increase in gastroenteritis burden with increasing temperatures (Bambrick et al. 2008). Similarly, general cases of salmonellosis were also found to be increased 5–10% with every degree increase in temperature in several European countries (Kovats et al. 2004; Tirado et al. 2010). Aflatoxin contamination of maize in African countries has also become a concern due to increased incidents between 2017 and 2022 which may be attributed to increasing temperatures and climate (Nji et al. 2022).

Climate change is found to affect animal susceptibility to diseases directly and indirectly. Conducive to pathogen proliferation lifestyle conditions can become part of animal's life due to climatic changes. Animal husbandry practices like use of veterinary drugs and antibiotics may change and affect food safety in return. Intense cold or drought conditions are found to increase the risk of bacterial diseases like mastitis in cattle which need the use of antibiotics. These antibiotics when found in meat negatively affect humankind also raising the issue of antibiotic resistance as Heat load was found to directly impact the increased prevalence of pathogens like *E. coli, S. uberis, and Staphylococcus aureus* along with increasing the incident of clinical mastitis in cows (Vitali et al. 2020). India-based study in which 12 years of data regarding mastitis and climatic conditions were gathered also inferred positive relation between hot weathers and incidence of mastitis in cattle (Jingar et al. 2014).

Extreme weather conditions like droughts and floods are related to soil contamination of substances that originally come from sewage, industrial, and agricultural places. Soil contamination of agricultural lands with dioxins and polychlorinated biphenyls (PCBs) are associated with inland floods due to climate change (Tirado et al. 2010). The Aral Sea area of Central Asia faced rapid desertification due to climate change and agricultural mismanagement which resulted in water, soil, and food contamination in that area (Muntean et al. 2003).

Ocean warming leads to acidification and affects water microflora, fisheries distribution, and algal blooms (Tirado et al. 2010). An increase in ocean temperature by one degree is found to increase 4.4% uptake of methyl mercury by fish, since elevated water temperature promotes mercury methylation (Booth and Zeller 2005). Natural disasters due to climate changes also cause emergency situations in which food safety risks arise (Tirado et al. 2010).

Changes in climate mean that food will be produced in different conditions where there will be a different use of pesticides, use of veterinary medicines, and emerging pathogens. Since most of these changes are unpredictable for now, it is need of the day to have quick and efficient surveillance mechanisms of food safety (Herrera et al. 2016). Artificial intelligence is being employed by researchers to predict changes in climatic factors and its impact on different aspects of food change that can challenge food safety measures. These models can further be upgraded to predict intensity and occurrence of microbial infestations and food safety risks based on the changes in weather parameters (Karanth et al. 2023).

9.4.3 Kitchen and Equipment Sanitation

Kitchen and equipment sanitation and cleanliness is also a very important factor affecting food safety. The kitchen is the place, and equipment is the tool that comes in direct contact with food. Hence, their contamination leads to food contamination and their unsafety affects food safety. In a study done on Korean kitchen conditions, 21 out of 50 kitchens were reported to have *S. aureus* on their dish clothes emphasizing the importance of kitchen sanitation for food safety (Hong and Lim 2015). The kitchen should be designed not to accumulate dirt, have proper ventilation, and should be made of material that is easy to clean. It is a difficult task to design such kitchens due to ergonomic conditions, cooking habits, and local trends. However, small changes in the kitchen can improve food safety conditions. In domestic settings, moving the sink (washing area) inside the kitchen can improve handwashing during food handling up to five times. Moreover, inclusion of sink in kitchen design was also found to lessen the risk of cross-contamination and increased frequency of rinsing actions (up to 11 times). The concept of food safety triangle with working area (kitchen table), washing area (sink) and cooking area (stove) can also be employed in new kitchen to improve food safety practices (Mihalache et al. 2022). Similarly, equipment should be cleaned and used with care. It should be made up of material that is noncorrosive, nontoxic, and easy to disinfect (Ucar et al. 2016).

9.4.4 Personal Hygiene of Food Handlers

Humans are the basic source of contamination and act as carriers. Food handlers and persons involved in food safety pose the greatest risk. It is estimated that 25–40% of all foodborne diseases are caused due to lack of hygiene by food industry handlers (Ucar et al. 2016). It was estimated that 93% outbreaks among 81 cases, collected from published articles between 1975 and 1995, involved food handlers (Guzewich and Ross 1999). Similarly, Hardstaff et al. (2018) also found 46% positive samples among food handlers in norovirus outbreaks in different food system settings.

Hence, personal hygiene is a huge matter of concern for food safety. It is important that the personal hygiene principle should be followed by workers in the food industry. Diseased individuals should not be hired in food industry which pose a risk to contaminate food, healthy individuals should have regular medical checkups, and any worker falling ill or having any injury should be removed from the food handling area and allowed to rest on leave (Ucar et al. 2016).

9.4.5 Globalization in Food Industry

With the recent advances in technology and methods of transportation, the world has become a global village. No doubt globalization has benefited nations in various ways, and at the same time it has posed some challenges too in front of governments, organizations, and individuals who are trying to ensure food safety. To save labor and material costs, food industries across the globe tend to import ingredients from across the world. These ingredients travel thousands of kilometers allowing food safety hazards of one place to affect the people at the other place. In other words, contamination that occurred in one country may affect the health of people in another country located on the opposite side of the globe. This means that governments not only have to deal with local contaminants, but they also must think about contaminants found around the globe (Fukuda 2015). The food transition that has occurred in developed countries is now taking place in developing countries but at a faster pace. Traditional food systems use simple methods, and hence food safety management is comparatively easy.

However, with the shift toward modernization and westernization, simple food items get replaced by processed food making food chains more complex, sophisticated, and long. This demands a lot of resources in terms of money, technology, and skills that pose great challenge to underdeveloped and developing countries (Guyomard et al. 2012). Increased consumption of perishable foods is observed with increase in incomes, that need more effort in storing, processing, and transporting. Similarly, advanced stages of transition mean complex supplier connections with food services generating more variety of sources from which food hazards can leach into the food systems. The increased food trade between countries at different stages of food transition pose challenges in terms of food safety for both sides (Unneverhr 2022). Moreover, food-related laws and regulations also differ from place to place. An ingredient legal at one place may be illegal at the other. Globalization has made it difficult to control foodborne diseases (infectious and noninfectious, both) (Hoffmann and Harder 2012). An example of this is the incident of 1996 when Guatemalan raspberries caused an outbreak in the United States (Calvin et al. 2003). Similarly, in the winter of 2008–2009, *Salmonella* that originally contaminated a Georgian Peanut processor caused illness among 714 people around 46 states (CDC 2009). Globalization has promoted urbanization in some countries which brought sanitation and cleanliness problems, affecting the food industry as well. Moreover, as food production shifts to the industrial level, a lack of trained staff following all the hygienic practices also becomes a problem. The problem of adulteration at the international level is also a major consequence of globalization (Hoffmann and Harder 2012).

To solve food safety challenges posed by globalization, collaboration at international and multisector levels is required. Collaborative and coordinated actions based on proper research across the globe should also be encouraged. International trade regulations should also be revised keeping in mind food safety. It is needed that leaders and researchers around the globe sit together to find proper solutions to problems faced by the world. Currently, the World Health Organization in collaboration with the Food and Agriculture Organization (FAO) is trying to bring stakeholders to one place and harmonize food systems by doing things like Codex Alimentarius Commission that is jointly managed (Fukuda 2015).

9.4.6 Food Adulteration

Food adulteration can be defined as the deliberate addition of compounds to food that are low quality, toxic, or nonedible to gain economic benefits (Kamruzzaman 2016). Food adulterants are neither contaminants that accidentally come from the environment nor anything that is added during any process across the supply chain but are intentionally added usually for economic gains (Spink et al. 2019). Food adulteration is not a new problem and has existed since the human history of food selling and buying exists, but unfortunately it manifests itself each time in a new way. Over the course of years, there have been many food adulteration incidents recorded which harmed not only people's health but also costed some lives. One of these incidents is the use of melamine in infant formula milk by Chinese industrialists to artificially increase protein content that affected 300,000 infants and even killed 6 infants according to some reports (BBC 2010). The actual scale of adulteration cannot be calculated since most of the cases go unnoticed or unreported. This is due to the inability of consumers to judge purity, adulterants being difficult to differentiate, difficulty in sampling every produce for adulteration due to increased number of products and economic issues, and difficulty in traceability of adulteration source due to hypercomplexity of food chains (Kamruzzaman 2016). Though Grocery Manufacturers Association in 2010 estimated that food frauds globally cost the food industry 10–15 billion dollars every year (Johnson 2014).

With the recent advances in technologies, procedures to fight adulteration have been developed, but at the same time, fraudulents have developed new techniques to deceive customers and authorities. The problem of adulteration is still a major challenge to food safety across the globe and continues to be widespread in both developed and nondeveloped countries. Hence, there is need to research and develop efficient, cheap, and quick techniques to detect adulteration (Kamruzzaman 2016). Moreover, policies and laws concerning adulteration should also be updated from time to time. Public–private partnership between the government and industry has recently been studied and employed as an effective and efficient way to deal with adulteration. Food fraud screening as well as vulnerability assessment can also help in reducing the intensity and frequency of adulteration incidents (Spink et al. 2016). Employing block chain technology in food supply chain can also help in recording information like conditions, shelf time, process, equipment, producer, and distributer, which can help the government to track the person involved in adulteration easily (Valoppi et al. 2021).

9.4.7 Misuse of Food Additives

Food additives are substances that are added to food (directly or indirectly) for their intended use (FDA 2023). Even though food additives are permissible to use, some of them are permissible to use up to some allowed limit.

The use of food additives above this limit may pose risks to human health. Unfortunately, several cases have been reported where food additives were either used above the prescribed limit or the use of nonpermissible food additives was found to be a problem to food safety (Gizaw 2019). A study done in India showed that 31% of the products chosen for sampling had nonpermissible colors in them, while 8–20 times above their approved limit were permissible colors found in crushed ice (Tripathi et al. 2007). Similarly in Pakistan, a study done showed that 4% of branded and 30% of non-branded beverages were unfit for human consumption for the same reasons (Saleem and Umar 2013). These food additives may pose serious health problems like cancer. Hence, controlling their use in food is still a big challenge for food safety.

9.4.8 Use of Products past Their Expiry Dates

"Expiry date" and "use before" dates are put on food labels to mention the date before which the food can be consumed safely without any risks (EU 2015). Foods after the mentioned dates should not be used, but unfortunately many markets do not remove expired food from their shelves, and consumers buy these foods and consume them exposing themselves to food safety risks. Moreover, it is observed that many markets put their near-to-expiry items on sales. Consumers buy these products in bulk without looking at expiry dates and consume these expired foods neglecting the purpose of these expiry dates. In an online survey done in Canada, 82% of respondents stated that their stores often sold expired items (Burnett et al. 2015). Moreover, another study in Kenya showed that consumers of vulnerable groups kept using expired food items for days after their expiry (Kunyanga et al. 2011).

To prevent food safety risks that can happen due to eating expired foods, governments should make policies and encourage markets to discard their expired foods.

9.5 Food Safety at Different Stages

9.5.1 Food Safety at Production Stage

With the increase in the human population, the need to increase world food production accordingly has led to the intensive use of agrochemicals in agriculture. These agrochemicals have no doubt increased production and protected crops but at the same time have posed great threats to human health and food safety. Chemicals like additives, pesticides, fertilizers, hormones, and antibiotics that are used in food production pose serious health hazards to humans. It is also said that fresh agricultural products are the source of certain illnesses in the United States (Rezaei 2018).

The use of fertilizers and pesticides is an important reason for food contamination. Pesticides are compounds that are used to kill pests that are harmful to crops. Some studies have detected pesticide residues in fruits and vegetables. Moreover, some derivatives from pesticides are harmful to human health (Kamboj et al. 2020). The US Environmental Protection Agency reported based on animal studies examining 321 pesticides, 146 out of them could possibly act as carcinogens. Pesticides not only contaminate crops and plants-based foods, but they also can contaminate animal-based foods and render them unsafe (Bhandari 2003). Organochlorine compounds, a type of pesticide, have also been detected in fatty foods (Kamboj et al. 2020). WHO (2016) suggests that 19% of all the cancers can be attributed to pesticide exposure. Harmful effects of pesticides have urged the world to either ban them or set maximum residue limits for them. Unfortunately, in many parts of the world, pesticides are still being used in ample amounts, e.g. China where an estimated 1.41 million tons of pesticide was reported to be used in 2018 (Li et al. 2021). The impact of these pesticides depends on their type like organophosphates, and carbates induce acute intoxication while organochlorines are carcinogenic (Gomes et al. 2020).

Heavy metals are an important risk to food safety and soil contamination with industrial waste being an important source of metals in food (Krishna and Govil 2007). Moreover, plants can easily absorb other toxic materials from soil and water, causing contamination not only of fruits and vegetables but of animal products and seafood too. A study in 2011 showed that 10% of all the rice produced in China had Cadmium in it which is a toxic heavy metal. Pest and insect infection and nonobservance of harvest principle also pose risks to food safety. These points also act as an important factor in increasing the risk of aflatoxin (cancer-causing mycotoxins) contamination (Rezaei, 2018).

Lack of resources, knowledge about food safety, farmer training programs, risk assessment, and control programs like HACCP at agriculture and production stage are the major problems in the way of food safety at this level. Farmers should not only be trained to ensure food safety at the production level, but collaborative research and monitoring of food production should also be done (Razaei 2018). It has been observed that farmers tend to adopt food safety practices more when they self-consume their produce as compared to those who only sell their produce (Hoffmann and Jones 2021; Kadjo et al. 2020). That's why apart from education and awareness of farmers, some economic incentives like subsidies over food safety technologies and price premium upon food safety have been proved beneficial strategies, especially in developing countries where farmers are sensitive to prices and inflation. Technology incentives as well as subsidies for reducing aflatoxins in stored maize were given in Kenya to compare their impact. It was observed that farmers were more inclined to use less costly food safety strategies like using plastic barrier as compared to costly technologies like mobile dryer with an increased use in plastic barrier (3–47%) when provided for free by the government (Hoffmann and Jones 2021). Enforcement of quality standards and guidelines has also been found beneficial in motivating farmers and producers to follow food safety guidelines as observed among Senegalese onion producers when certifying quality labels were enforced (Bernard et al. 2017).

9.5.2 Food Safety at Processing and Packaging Stage

At the processing and packaging stages, some processes generate compounds sometimes that act as food contaminant and render food unsafe for human consumption. Heating is a common process that is done in food industries. External factors in combination with high cooking temperature generate toxic compounds. Similarly, other food processing procedures like canning, baking, heating, grilling, and roasting may generate toxic compounds like acrylamide, polycyclic aromatic hydrocarbons, nitrosamines, and furans if not carried out properly (Nerin et al. 2016). Moreover, the use of disinfectants and sanitizers in amounts above the prescribed limit is also known to pose risks to food safety by accumulating in food (Kamboj et al. 2020). Although different laws and regulations to ensure food safety have been made and implemented by national and international bodies, different accidents due to processing errors continue to hamper food safety imparting human and economic losses. *Salmonella* contamination in an international chocolate producing industry in 2022 affected 450 individuals around the globe in 2022 (FSN 2022). Recently, Abbott Nutrition had to recall *Cronobacter*-infected infant formula from the market of three American states after two infants died. This incidence was a proof that the risk of food safety is always there, though its impact on human health can be reduced by immediate and active action (CDC 2022).

Food is packaged to protect it from physical injuries and contamination. It increases the shelf life of food, but unfortunately the material used for food packaging sometimes acts as a source of food contamination. For example, the internal part of the can is coated with chemicals like epoxy resins to prevent corrosion, but by-products of these resins like bisphenol

A, cyclo-di, and bisphenol A diglycidyl ether (BADGE) can contaminate food acting as endocrine disruptors (Cabado et al. 2008). Paper and board are also one of the most used food packaging materials, and contamination of food with printing inks is one of the common packaging problems to food safety (Kamboj et al. 2020).

Moreover, cross-contamination of food due to improper handling, lack of sanitation, and personal hygiene of food handlers is one of the major causes of food-linked illnesses. Thus, proper conduction of food safety systems, proper training, proper monitoring of food handling, proper sanitation, proper testing of packaging materials for contamination before use, etc., are some of the steps that should be implied to prevent food safety-related risks in the industry.

9.5.3 Food Safety at Transport and Distribution Stage

After the food products are processed or when food commodities are produced at agricultural lands, they are transported to different places where they are offered for sale. Food contamination can occur during this stage as well. Moreover, since one truck can be used for transportation of different kinds of products, the chances of cross-contamination at this stage also exist. Cross-contamination of food with chemicals used for disinfection or other sources is also a big risk to food safety. Moreover, for long-distance transportation, high barriers are used for protection which are not tested for their ability to protect against food contamination (Rather et al. 2017). In 1999, the European Economic Community found that contaminated pallets with fungicide were used for food transportation and storage which caused disease in humans (Kamboj et al. 2020). Food contamination during storage is also an issue. Lack of proper storage conditions deteriorates food safety and quality. Factors like direct sunlight, moisture, and temperature affect food deterioration and may cause the generation of toxic materials from the packaging. Food packaging that exhibits proper barrier properties during long-term storage should be used. Perishable food like dairy, eggs, and meat should be stored under cold temperatures (Nerin et al. 2007).

9.5.4 Food Safety at Consumption Stage

Food safety in the household is a very important factor in ensuring the health of people around the globe. Even if the food is produced, processed, or packaged safely but is not cooked, stored, or handled properly at home, it will become unsafe and cause diseases. Improper food management at consumption stage is a big challenge to food safety both in developed and developing countries. According to WHO, 40% of the foodborne illnesses in Europe are caused by food that is consumed in homes (WHO 2003).

In 2015, a literature review was done under USDA. It showed that the most common reasons for foodborne illnesses and contamination in households include use of unsafe water, lack of sanitation practices, reheating of food at inadequate temperature, and inadequate cooking and storage temperatures of foods. Not washing hands before and during food preparation, lack of personal hygiene, and the use of unclean utensils are also major reasons for food contamination at homes (Woldt and Moy 2015). A study in 2012 also reported that direct contact between raw and cooked food was avoided by only 36.5% of consumers (Langiano et al. 2012).

The use of microwaves is very common in households today and poses a threat to food safety sometimes. Food is heated in some containers in microwaves. Microwavable materials like plastics, composites, and paperboard, when microwaved, produce compounds that get transferred into food, thus reducing the safety of food. Disinfectants used in home kitchens and the food industry also act as a food safety hazard when they enter the food as residues (Rather et al. 2017).

WHO advises five acts to ensure safer foods at home which include cleaning, keeping raw and cooked foods separate, proper cooking of food, adequate storage temperatures, and use of safe raw materials and water (WHO 2006). Proper awareness and knowledge of safe food handling and preparation should be given to the public with the help of multisector collaboration at the national and international level (Figure 9.1).

9.6 Sanitation and Personal Hygiene

9.6.1 Sanitation

Sanitation refers to public health conditions and health efforts to maintain and provide better health to people by maintaining cleanliness. Melosi provides sanitation examples like providing clean water for washing hands and garbage bins (trash cans) to avoid carelessly throwing of food (Melosi 2004). Food can become unhygienic and unfit for consumption at any stage during processing, preparation, and transportation. Sanitation is a prevention effort, process, or action that is required to keep the food free from hazards that cause health-damaging effects to food (Rahman et al. 2018). Leading factors of foodborne illness are improper storage and reheating that contribute 50%,

Figure 9.1 Food safety and sanitation.

inappropriately stored food contributing 45% and cross-contamination contributing 39% foodborne illnesses (Kamboj et al. 2020). More than 200 known diseases are transmitted through food (Mead et al. 1999). Foodborne illness can be caused by contaminants, improper food handling practices, and food allergies (Victoria 2015). Poor sanitation not only causes foodborne illness but also water-related illnesses. In 2002, Schistosomiasis infected nearly 200 million people in 70 countries (Gleick 2002). Water is used in the production of food, in cleaning freshly cut food, in the preparation of food, and for drinking purposes. Where water is used for reducing potential concerns of pathogens, it also transfers pathogens to food (Gil et al. 2009). As proposed by American Dietetic Association, the only solution to the problem of unsanitary water and food requires a multidisciplinary approach that includes experts of sanitation, public health, agriculture, epidemiology, water resources, and nutrition (Albercht et al. 2009).

People assume that sanitation is important just because they were told and do not know about the reason why it is important. Brackett explains there are three reasons why the producers should be concerned with sanitation and personal hygiene: necessity, quality, and safety. Quality and safety of products are dependent on their microbiological flora. In terms of microbiome, specific standards are produced based on consumer preferences and specific conditions like for curd and other prebiotics products. Due to unhygienic and unsanitary conditions, producers are at risk of exceeding these specific standards. From production to consumption, every step affects the quality of food. In the end, contaminated food spreads foodborne illness (Brackett 1992). As for dairy products, pathogens are found frequently in the initial raw milk (Oliver et al. 2009). Therefore, it is necessary to provide sanitary conditions to food completely to reduce pathogens such as *Listeria monocytogenes* (Campdepadros et al. 2012).

9.6.2 Personal Hygiene

Personal hygiene means an individual's personal cleanliness. Individual's body parts should be kept clean to break off the transmission of diseases. It also gives a good esthetic value to others. Science is also focusing on personal hygiene and sanitation to provide better public health and environment (Figure 9.2).

In Ethiopia, 80% of diseases are caused by poor personal hygiene and poor environment. To improve the quality of life, personal hygiene and sanitation should be encouraged as poor personal hygiene and sanitary conditions are causing hazardous diseases, and education about personal hygiene is needed to be spread widely. Disease transmission requires a source of infection, a route of transmission (host), and an environment. Hence without a host/route, spread of pathogens is not possible (Legesse and Ambelu 2004). Personal hygiene determines the quality of food from handwashing to cleaning of plates and removing damaged parts of food to enhance the integrity of food (Clyton et al. 2002). Workers should keep in check their personal hygiene. The following are some things that should be kept in check (Table 9.1).

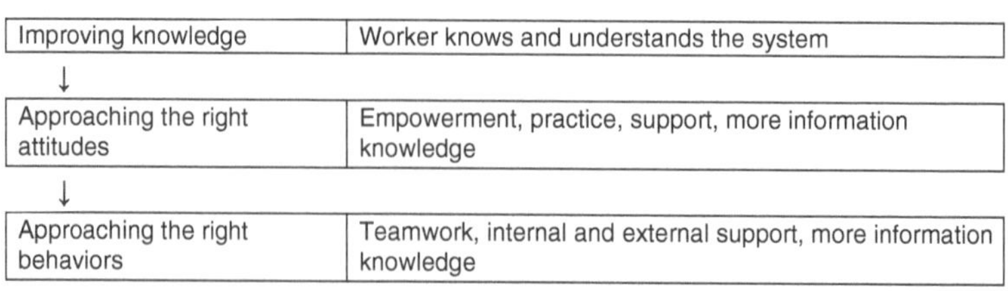

Figure 9.2 Personal hygiene pathways. *Source:* Gilling et al. (2001).

Table 9.1 Personal hygiene checklist.

√	**Personal hygiene**
√	Hair should be properly tucked inside the cap
√	No earning or necklace/chains
√	No outer pockets
√	Wear neat and clean clothes
√	No wristwatch/rings
√	Cover all wounds
√	Nails should be sort and clean
√	Torn clothes should be repaired or replaced
√	Wear clogs and safety shoes

Dust and dirt are everywhere in the environment. Individuals are in constant contact with it. Besides this, we sweat also and secrete oil from our body parts including our hands. The very first thing that affects the environment is our skin and our hands. If we do not wash our hands and continue our work, we are helping pathogens to grow and spread in a healthy environment. By simply washing our hands, we can save ourselves and others from getting ill.

On the other hand, if we do not consider personal hygiene, it will lead to skin diseases and infections, and intestinal worms' accumulation in nails by touching food during preparation, cooking, and eating, which will rest in mouths causing many severe foodborne diseases. Any body part if left dirty, will be home to pathogens, creating foul smell and weakening the immune system of body, making one more vulnerable to diseases (Legesse and Ambelu 2004).

9.7 World Condition of Sanitation

Provision of safe and healthy food and water is a major public health issue and always has been a nutritional concern (Sobsey 2006). In 1999, nearly 747 million people had sanitation facilities. In 2005, about 60% of people were deprived of sanitation facilities; nearly, 2.4 billion people had no access to hygiene facilities and sanitation (WHO 2005). There are 4.5 billion people lacking sanitary conditions, while 2.3 billion people do not even have basic sanitation facilities. Then, 892 million people defecate openly in rural areas.

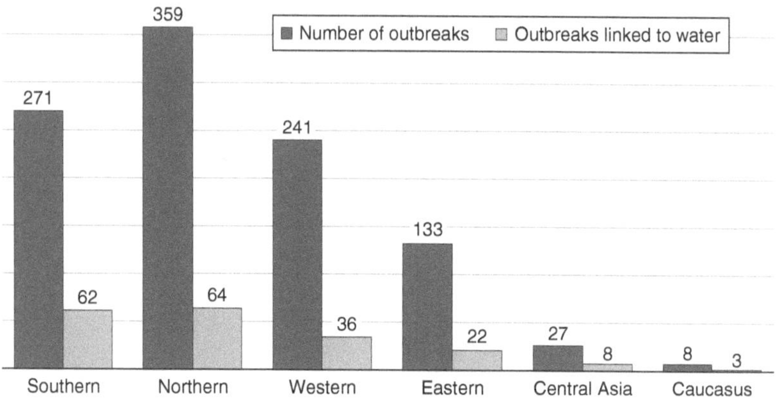

Figure 9.3 Disease burden in pan-European region (2010–2013)
Source: Kulinkina et al. (2016).

To eliminate the global sanitation crisis, WHO established "World Toilet Day" in 2001. Narendra Modi launched a project Swachh Bharat Mission (Clean India Mission) in 2014 (Zhou et al. 2018). In 2015, China proposed "Toilet Revolution" in rural areas to elevate sanitation (Cheng et al. 2018). More than 68,000 public toilets were restored in China (Wang et al. 2018). According to WHO, the percentage of open defecation decreased from 24% to 13%, and sanitation improved from 54% to 68% (WHO 2015).

According to WHO, 785 million people lack access to clean water for drinking and 2 billion people are using drinking water from feces-contaminated sources. About 3 billion people lack basic handwashing facilities at home. In 2016, about 1.9 million deaths and 12 million disability-adjusted lifestyles (DALYs) could be prevented by WASH. Safe water, sanitation, and hygiene are collectively called WASH (WHO 2019). Contaminated and unhygienic water in food becomes the source of infection and causes foodborne illness (Marino 2007) (Figure 9.3).

Each year 1 billion humans are infected by diarrheal disease and 2–5 million people die by diarrheal disease. Other water-related diseases are intestinal helminths, schistosomiasis, dracunuliasis, tracoma, poliomyelitis, and trypanosomiaisis (Gleick 2002). There are other diseases that are not caused by pathogens but by chemicals in water. Around 140 million people in 70 countries are drinking water with excessive arsenic (WHO 2018), and this can lead to skin lesions and cancer, cardiovascular diseases, diabetes, impaired cognitive development, and deaths. Preventive measures and all different systems to prevent diseases due to poor sanitary conditions are quite expensive. WASH contributes to economic well-being and is helpful in improving water sanitary conditions. According to WHO, every dollar invested in water sanitation returns US$5, an estimated gain of 1.5% of global gross domestic product (WHO 2019).

9.8 Foodborne Illness

As the trend of dining out is increasing, the number of cases of foodborne illness is also increasing. The contributory reason is lack of knowledge on personal hygiene and food hygiene awareness and its implementation (Kamboj et al. 2020). Foodborne illness is a public health issue that is common and sometimes life-threatening, yet it is preventable. In United States, about a total of 48 million illnesses, 128,000 hospitalizations, and 3,000 deaths occur each year because of foodborne illness as a major health problem (Scallan et al. 2011). Among all foodborne disease-causing bacteria and viruses (pathogens), *Salmonella, Listeria*, *E. coli*, norovirus, and hepatitis A are the most common.

Workers related to food industries require knowledge about handwashing, hygienic practices, and personal health to provide safe food (Bas et al. 2006) and protect it from contaminants causing diseases. Keeping in mind the reasons of foodborne diseases mentioned in Table 9.2, we can understand why Langhar also emphasizes that changes in required microbiological flora cause foodborne illness. According to Langhar, nonpasteurized dairy product consumption increases the risk of diseases by 150 times than pasteurized product consumption (Langhar et al. 2012). Without the knowledge of sanitation and personal hygiene, food workers can be a major source of contamination leading to different diseases (Clyton et al. 2002) (Table 9.3).

Table 9.2 Sources of foodborne diseases.

Pathogenic organisms of concerns	Pathogenic organisms of potential concerns
Listeria monocytogenes	*E. coli* O157:H7
Clostridium botulinum	*Aeromonas hydrophila*
Shigella sp.	*Bacillus cereus*
Salmonella sp.	*Plesiomonas shigelloides*
Parasites and viruses	*Yersinia enterocolitia*

Source: Brackett (1992).

Table 9.3 List of pathogen, sources, diseases, and their prevention.

Pathogen	Source	Disease	Prevention
E. coli	Contaminated food and water	Severe illnesses	Safe food handling, cooking meats thoroughly, and avoiding unpasteurized milk and juices
Norovirus	Contaminated food and water. Person-to-person contact	Diarrhea, cramps, headache, mild fever, and muscle ache	Safe food handling and good handwashing
Salmonella	Contaminated food and water	Salmonellosis Enteric fever	Safe food handling, good handwashing, cooking meats thoroughly, separate raw and cooked foods, and keeping food at the correct temperature
Campylobacter	Contaminated food, raw milk, and cheese, raw meat, and poultry	Sporadic nausea, vomiting, and diarrhea (sometimes bloody)	Safe food handling, good handwashing, cooking meats thoroughly, and avoiding raw milk and cheese
Hepatitis A virus	Contaminated food and water, shellfish, and salads	Hepatitis A	Good handwashing
Listeria	Contaminated food and water.	Invasive listeriosis, nausea, vomiting, headache, mild fever, and sometimes diarrhea,	Good handwashing, keeping the refrigerator clean and at 40 °F, separating raw and cooked foods, and avoiding raw milk and cheese.

Source: FDA (2020).

9.9 Sanitation at Industrial Level

Nowadays, people are busier and intend to dine out or tend to have prepared foods. In the United States, $997 billion sales is projected in the restaurant industry in 2023 and 15.5 million people are employed in the food industry

(NRA 2023). Water is a necessity, and it is necessary for almost every step in the food processing industry. Either in washing raw food, cooking ingredients, or in cooling of cooked foods. This gives a clear picture that the amount of water used in food processing and catering is quite large and the risk of contamination increases using potentially unhygienic water. Water is also used in cleaning premises and equipment. Maintaining the quality of water requires close monitoring to eliminate every single chance of contamination. Certain technologies like flat belts, wire belts, and metal rollers have some weaknesses in providing efficient cleaning. Modular plastic proves better than others. It reduces debris and provides better clearing facilities in the conveyer belt (Scheffler 2009). Certain chemicals are used in food processing industries for cleaning purposes and others, and some of them are listed down (Table 9.4).

These chemicals are added to water at certain concentrations and at certain temperatures and then used for certain purposes. Specific monitoring is required; otherwise, concentration may contaminate water and can cause harmful effects. There are certain regulations and standards of water supplies in England and Whales. "The Water Industry Act 1991," "The Water Act 2003," and "The Security and Emergency Measures Direction 2006" are for the provision of safe and quality water for drinking purposes and in food processing. This also includes inspection of water and disinfection systems like filtration (RSPH/DWI/Water 2017).

Food workers can spread 10,000–100,000 microorganisms in every minute (Ucar et al. 2016). To maintain sanitation and personal hygiene, workers should be empowered with the right knowledge and proper practice of that knowledge to create a safe environment for food production. Before employment, every worker should undergo health examination and on daily basis, SOPs should be followed (ASEAN 2015). Food hygiene is an action that provides food safety from food production to consumption. At any point in the food supply chain, food can be contaminated and become unhygienic for consumption (Kamboj et al. 2020).

Table 9.4 Cleaning chemicals.

Chemical	Disadvantages	Chlorine
Corrosive and irritating to skin	Iodine	Discoloration of equipment and surfaces
Quaternary ammonium compounds	Not compatible with hard water and detergents	

Source: CFSPH (2023).

To maintain food safety and hygiene, there are certain food safety systems like Good Manufacturing Practices (GMP), Sanitation Standard Operating Procedure (SSOP), Good Hygiene Practices (GHP), and Hazard Analysis of Critical Control Points (HACCP) (Kamboj et al. 2020). Besides the personal hygiene of workers, industrial equipment and premises should also be sanitized and cleaned thoroughly.

- Good Manufacturing Practices (GMP): GMP regulations were introduced in 1969 by the Food and Drug Administration (FDA) and revised in 1986. GMP provides requirements for food and drug production in sanitary conditions. The FDA developed GMP for everything except meat, poultry, and eggs. GMP regulations include general maintenance of physical facilities, cleaning, sanitizing, storage and handling of utensils and equipment, pest control, employee training, and practice. Buildings, facilities, plant design, production and process controls, and quality assurance assessment are also parts of GMP.

  Food handlers are the source of cross-contamination; therefore, training in food handling and personal hygiene is given to food handlers. Regular checkups are held, and proper monitoring is done to ensure food safety (Keener 2009).
- Sanitation Standard Operating Procedures (SSOPs): SSOPs are daily routine procedures to carry out standard sanitation. They are specific steps to sanitation and prevent product adulteration. US Department of Agriculture of Agriculture's Food Safety Inspection has regulations for meat, poultry, and eggs. SSOPs are utilized in the production of these products. SSOPs are signed and dated by the plant owner and monitored periodically (Keener 2009). It ensures good sanitation, assists with consistency, makes training easy for employees, and assists with quality. SSOP has six main components:

  1) General rules
  2) Development of SSOP
  3) Implementation and monitoring of SSOP
  4) Maintenance of SSOP
  5) Corrective actions
  6) Record keeping

SSOP plans are made to describe sanitation procedures to be used in plants. It is a written set of sanitation procedures, and every employee uses this as a tool for training. SSOP acts as a base for monitoring programs that management and workers all are following sanitation requirements. SSOPs identify trends and prevent the recurrence of mistakes by making corrections (Gok and Kayaardi 2000).

Assemble team
↓
Describe product
↓
Identify intended use
↓
Construct flow diagram
↓
Identify hazard analysis accessing less risk
↓
Identify Critical Control Points
↓
Establishing Critical Limits
↓
Implementing continuous monitoring
↓
Implementing corrective measures
↓
Verification and validation
↓
Documentation and record-keeping

Figure 9.4 SSOP plan pathways. *Source:* Ibrahim (2020).

- Hazard Analysis and Critical Control Point (HACCP): HACCP was developed in 1960, and developed HACCP system was presented in 1971. HACCP is a management system through which food safety is maintained by eliminating foreign substances that are hazardous in raw material or produced at any point during processing, distribution, retail, and food consumption. HACCP identifies and assesses hazards associated with food, determines CCPs (critical control points) to control health hazards in food processing, and develops systems to monitor CCPs (Figure 9.4).

9.10 Sanitation at Household Level

Food is termed as safe, when it does not lose its nutritious content and freshness and when it is clean in physical, chemical, and microbiological terms (Ucar et al. 2016). Poor sanitation can result in the reduced shelf life of foods and reduced quality of food, overall causing foodborne illness to the consumer. Sanitary regulations are not only for restaurants, catering services, or food processing industries; these food safety regulations should also be applied at household level. The public should also be aware of sanitation and personal hygiene to avoid diseases and promote better health (Janjic et al. 2017).

The hygiene quality of food is decreased by purchasing stale food, storing it in inappropriate conditions, cooking in large amounts than required, eating it with dirty and unhygienic hands, and storing raw and cooked food together. Improper heating, cooking, and cooling of food cause 44% of foodborne illnesses. Poor hygiene and reuse of leftovers cause 14% of other diseases. There are certain allergens that should be separated from other food substances to avoid cross-contamination. Twenty-five percent of reported foodborne illnesses are caused by improper handling of food by the consumer and improper ways of cooking food in the household (McCabe-Sellers and Beattie 2004).

The information about cross-contamination should be delivered to consumers explaining What is cross-contamination, How it occurs, Where it occurs, and What can be done to prevent cross-contamination. In Eastern Sudan, the source water was in good condition microbiologically. When it was transported via donkeys to home, its quality deteriorated as the same may happen to our groceries (Legesse and Ambelu 2004). Home is a place that is continuously in contact with different pathogens in one way or another. The source of pathogen can be people, shoes, pets, and food. Nowadays, coronavirus has also focused household and individuals' attention toward sanitized, hygienic practices, and proper handwashing. Improper cleaning of utensils and crockery to avoid the production of biofilms prevents the sanitation of utensils and causes accumulation of bacteria on surfaces of utensils. A kitchen is the most critical place in a house where pathogens usually survive. The importance of personal hygiene is acknowledged in every step of the food chain, but the last component of food chain consumers is studied least and needs attention the most.

The information on households is received from two sources: foodborne disease outbreaks analysis and research-based consumer studies. Analysis of outbreaks does not provide us enough information about consumer's behavior concerning food safety and sanitation (Janjic et al. 2017). The main purpose of researching consumers is to study what percentage of foodborne illnesses are in households and how much consumers know about food safety and sanitation practices (Tucker et al. 2006).

Consumers are a very important part of the food chain and play an important part in preventing foodborne illnesses. They are the "Last line of defense" in food chain. Global food security will only be achieved when all components work together. Consumers should be aware of the risks of improper handling of food and poor hygienic practices to ensure food safety and prevent foodborne illness (Raspor and Jevsik 2008).

9.11 The Impact of Sanitation on Food Security

Water serves as a medium of absorbing various nutrients and improves biological utilization of food. We do not just use water for drinking purposes, but also it is needed in every step of food security. If it is unsanitary, it will arise food security and food safety issues (Miller et al. 2021). From food security definition, food availability lays the base of food security because if food is not available how can we eradicate hunger and malnutrition issues. Food availability also includes the availability of safe drinking water and sanitation services. Safe drinking water, sanitation, and hygiene are now recognized as WASH. WASH explains that WASH services are directly linked to food security (WASH 2013). Improper sanitation systems and water pollution lead to the irrigation of crops with water infested with pathogens, heavy metals, and pesticides which not only may disrupt crop growth but is a food safety risk consequently posing negative impact on food security. High concentrations of nitrogen and phosphorus in wastewater lead to algal blooming with cyanobacteria releasing cyanotoxins. When this cyanotoxin containing water is used for irrigation, it causes growth inhibition in crops (Malakar et al. 2019). Poor water quality also negatively impacts the food availability and safety in livestock where water is used for drinking. Toxins and heavy metals in polluted drinking water are known to cause problems in animal health and productivity which are further exacerbated by forage irrigated with the same polluted water (Linderhof et al. 2021).

Another element of food security is food accessibility. Lack of water and sanitation especially at household level leads to decline in food accessibility because most of their budget is utilized in securing safe water for their families and spending less on their food and nutrition which make nutritious food inaccessible. Food utilization is another important element of food security. If we do not utilize food at its maximum by our body through digestion, assimilation, and absorption, we are not getting any benefit of food security. Even if we have sufficient food, without safe drinking water, sanitation, and hygiene practices, many infectious diseases like diarrhea arise which compromise food absorption that leads to undernutrition. Diarrhea also makes your body defenseless and makes it more vulnerable to infections as it decreases resistance toward infections (WASH 2013). Water is used in food processing for many purposes. Poor sanitation practices and water quality lead to poor food safety that in turn negatively affects food security as mentioned earlier. Water being a major ingredient of some foods can destroy the safety of foods like use of cholera-infested ice in Jakarta leading to disease outbreak (Waturangi et al. 2013). A study based on the rural areas of west Cameroon showed multiple means through which poor water quality can impact food security. These included negative impact on food consumption due to lack of water available for preparing meal along with the use of poor-quality water in some cases which rendered food unsafe (Nounkeu and Dharod 2020).

In Agriculture sector, 43% of farmers are women who also spend six hours collecting water on daily basis to provide water to their families, which in result decrease their working time at farms (WASH 2013). This altogether affects the outcome of crops and food availability due to low manpower. Poor sanitation practices lead to a disease burden that is a major toll to economic conditions of households, thus hampering their ability to access enough food through their finances. Economic losses due to sanitation cost 6.4% of India's GDP (Singh 2014). Food spoilage and contamination due to poor sanitary practices, if avoided, can also help reduce the food waste that can make more food available for consumption. Hence, even though sanitation and food security are two distinct problems, they are interconnected and must be targeted simultaneously (Miller et al. 2021).

9.12 Conclusion

Safe food is a basic human right (UN 1948) which means that food should be free of any hazards and contaminants that pose a risk to human health. With the advances in technology and globalization, where food safety systems and various technologies to ensure food safety came into existence, at the same time new challenges and problems for food safety also arose like climate change, new adulterants, and use of additives. Food safety requires efforts at every step from farm to plate. Sanitation and hygiene are the most important factors that have a great impact on food safety. Food safety without sanitation is not possible. Hence, principles of sanitation and hygiene should be followed by everyone related to the food supply chain either as producer or handler or consumer. International organizations, governments, and stakeholders in the food industry should join hands to promote research, laws, awareness, and knowledge regarding food safety and sanitation so that the global burden of foodborne diseases around the world can be decreased.

References

Albercht, J.A., Nero, D.N., and American Dietetic Association. (2009). Position of the American Dietetic Association: food and water safety. *Journal of American Dietetic Association.* 109 (8): 1449–1460.

ASEAN. (2015). Sanitation and hygiene. In: *Annex VIII – ASEAN Guidelines on Good Manufacturing Practice for Health Supplements*, 23–25. Bangkok, Thailand: The Association of Southeast Asian Nations.

Asin, S. and Catala, S.S. (1995). Development of *Trypanosoma cruzi* in Triatoma infestans:influence of temperature and blood consumption. *Journal of Parasitology.* 81 (1): 1–7.

Australian Institute of Food Safety. (2020). What are the different types of food contamination. https://www.foodsafety.com.au/faq/what-are-the-different-types-of-food-contamination (accessed 5 December 2020).

Avgoustaki, D.D. and Xydis, G. (2020). How energy innovation in indoor vertical farming can improve food security, sustainability, and food safety? *Advances in Food Security and Sustainability.* 5: 1–51.

Bambrick, H., Dear, K., Woodruff, R. et al. (2008). The impacts of climate change on three health outcomes: temperature-related mortality and hospitalisations, salmonellosis and other bacterial gastroenteritis, and population at risk from dengue. *Garnaut Climatic Change Review, Australia* 2–47.

Bas, M., Ersun, A.S., and Kivanc, G. (2006). The evaluation of food hygiene knowledge, attitudes, and practices of food handlers' in food businesses in Turkey. *Food Control.* 7 (4): 317–322.

BBC. (2010). China dairy products found tainted with melamine. https://www.bbc.com/news/10565838#:~:text=Chinese%20food%20safety%20officials%20have,the%20toxic%20industrial%20chemical%20melamine.&text=Test%20samples%20showed%20the%20milk,babies%20and%20made%20300%2C000%20ill. (accessed 06 November 2023).

Bernard, T., De Janvry, A., Mbaye, S., and Sadoulet, E. (2017). Expected product market reforms and technology adoption by Senegalese onion producers. *American Journal of Agricultural Economics.* 99 (4): 1096–1115.

Bhandari, S.D. (2003). Hazards resulting from environmental, industrial, and agricultural contaminants. In: *Food Safety Handbook* (ed. R.H. Schmidt and G.E. Rodrick), 291–321. New Jersey: John Wiley & Sons, Inc.

Booth, S. and Zeller, D. (2005). Mercury, food webs, and marine mammals: implications of diet and climate change for human health. *Environmental Health Perspectives.* 113 (5): 521–526.

Brackett, R.E. (1992). Shelf stability of fresh produce as influenced by sanitation and disinfection. *Journal of Food Protection.* 55 (10): 808–814.

Burnett, K., Skinner, K., and Leblanc, J. (2015). From food mail to nutrition north Canada: reconsidering federal food subsidy programs for northern Ontario. *Canadian food studies/ La Revue canadienne des études sur l'alimentation.* 2 (1): 141–156.

Calvin, L., Flores, L. and Foster, W. (2003). Case study Guatemalan raspberries and cyclospora. International Food Policy Research Institute. Washington D.C., USA. https://lib.icimod.org/api/files/66313233-1ff0-42d9-b134-eb9ec359b46d/4351.pdf. (accessed 06 November 2023).

Cabado, A.G., Aldea, S., Porro, C. et al. (2008). Migration of BADGE (bisphenol A diglycidyl-ether) and BFDGE (bisphenol F diglycidyl-ether) in canned seafood. *Food and Chemical Toxicology.* 46 (5): 1674–1680.

Campdepadros, M., Stchigel, A.M., Romeu, M. et al. (2012). Effectiveness of two sanitation procedures for decreasing the microbial contamination levels (including Listeria monocytogenes) on food contact and non-food contact surfaces in a dessert-processing factory. *Food Control.* 23 (1): 26–31.

CDC. (2009). Typhimurium infections linked to Peanut Butter. https://www.cdc.gov/salmonella/2009/peanut-butter-2008-2009.html. (accessed 06 November 2023).

CDC. (2022). Cronobacter and Powdered Infant Formula Investigation. https://www.cdc.gov/cronobacter/outbreaks/infant-formula.html (accessed 06 November 2023).

CFSPH. (2023). *Disinfection 101 Key Principles of Cleaning and Disinfection for Animal Settings.* Ames, Iowa: The Center of Food Security and Public Health, Iowa State University, College of Veterinary Medicine.

Checkley, W., Epstein, L.D., Gilman, R.H. et al. (2000). Effects of El Nino and ambient temperature on hospital admissions for diarrheal disease in Peruvian children. *The Lancet.* 335 (9202): 442–450.

Cheng, S., Li, Z., Uddin, S.M.N. et al. (2018). Toilet revolution in China. *Journal of Environmental Management.* 216: 347–356.

Clyton, D.A., Griffith, C.J., Price, P.E., and Peters, A.C. (2002). Food handlers' beliefs and self-reported practices. *International Journal of Environmental Health Research.* 12 (1): 25–39.

EU. (2015). Best before date labels: protecting consumer and limiting food waste. Briefing European Parliamentary Research Service (europa.eu) (accessed 06 November 2023)

FAO/WHO. (2009). *Codex alimentarius: Food Hygiene (Basic Texts)*, 4e. Rome, Italy: FAO/WHO.

FDA. (2020). Foodborne pathogens. https://www.fda.gov/food/outbreaks-foodborne-illness/foodborne-pathogens. (accessed 6 November 2023).

FDA. (2022). How to cut food waste and maintain food safety. https://www.fda.gov/food/consumers/

how-cut-food-waste-and-maintain-food-safety#:~:text=Between%2030%20and%2040%20percent,food%20per%20person%20per%20month. (accessed 17 May 2023).

FDA. (2023). Understanding how the FDA regulates food additives and GRAS ingredients. https://www.fda.gov/food/food-additives-and-gras-ingredients-information-consumers/understanding-how-fda-regulates-food-additives-and-gras-ingredients#:~:text=A%20food%20additive%20is%20defined,the%20characteristics%20of%20any%20food (accessed 6 November 2023).

FSN. (2022). From outbreaks to MDMA and chemicals: 2022 in review. https://www.foodsafetynews.com/2022/12/from-outbreaks-to-mdma-and-chemicals-2022-in-review/ (accessed 6 November 2023).

Fukuda, K. (2015). *Food Safety in a Globalized World*. Geneva, Switzerland: WHO.

Fung, F., Wang, H.-S., and Menon, S. (2018). Food safety in the 21st century. *Biomedical Journal*. 41 (2): 88–95.

Gebeyehu, D.T., East, L., Wark, S., and Islam, S. (2022). Impact of COVID-19 on the food security and identifying the compromised food security dimension: a systematic review protocol. *PLoS One*. 17: e0272859.

Gil, M.I., Selema, M.V., Galev, F.L., and Allende, A. (2009). Fresh-cut product sanitation and wash water disinfection: problems and solutions. *International Journal of Food Microbiology*. 134 (1–2): 37–45.

Gilling, S.J., Taylor, E.A., Kane, K., and Taylor, J.Z. (2001). Successful hazard analysis critical control point implementation in the United Kingdom: understanding the barriers through the use of a behavioral adherence model. *Journal of Food Protection*. 64 (5): 710–715.

Gizaw, Z. (2019). Public health risks related to food safety issues in the food market: a systematic literature review. *Environmental Health and Preventive Medicine*. 24 (1): 68.

Gleick, P.H. (2002). Dirty water: estimated deaths from water-related diseases 2000-2020. Pacific Institute Research Report.

Gok, V. and Kayaardi, S. (2000). *Sanitation Standard Operating Procedures (SSOP) for Meat Industry*. Ankara, Turkey: Blacksea and Asian Symposium on Food Technology.

Gomes, H.D.O., Menezes, J.M.C., da Costa, J.G.M. et al. (2020). A socio-environmental perspective on pesticide use and food production. *Ecotoxicology and Environmental Safety*. *197*: 110627.

Guyomard, H., Darcy-Vrillon, B., Esnouf, C. et al. (2012). Eating patterns and food systems: critical knowledge requirements for policy design and implementation. *Agriculture & Food Security*. 13: 1–21.

Guzewich, J. and Ross, M.P. (1999). *Evaluation of Risks Related to Microbial Contamination of Ready to Eat Foods by Food Preparation Workers and*

Effectiveness of Interventions to Minimize those Risks. FDA, Centre for Food Safety and Applied Nutrition.

Hardstaff, J.L., Clough, H.E., Lutje, V. et al. (2018). Foodborne and food-handler norovirus outbreaks: a systematic review. *Foodborne Pathogens and Disease.* 10: 589–597.

Herrera, M., Anadon, R., Iqbal, S.Z. et al. (2016). Climate change and food safety. In: *Food Safety: Basic Concepts, Recent Issues and Future Challenges* (ed. J. Selamat and S.Z. Iqbal), 149–160. Switzerland: Springer.

Hoffmann, S. and Harder, W. (2012). Food safety and risk governance in globalized markets. *Health Matrix: The Journal of Law-Medicine.* 20 (1): 5–54.

Hoffmann, V. and Jones, K. (2021). Improving food safety on the farm: experimental evidence from Kenya on incentives and subsidies for technology adoption. *World Development.* 143: 105406.

Hoffmann, S., Maculloch, B., and Batz, M. (2015). Economic burden of major foodborne illnesses acquired in the United States. *USDA: Economic Information Bulletin* 140.

Hoffmann, V., Moser, C., and Saak, A. (2019). Food safety in low and middle-income countries: the evidence through an economic lens. *World Development.* 123: 104611.

Hong, J. and Lim, S.-Y. (2015). Microbial contamination in Kitchen and Refrigerators of Korean households. *Journal of Food Hygiene and Safety.* 30 (4): 303–308.

Ibrahim, O.O. (2020). Introduction to Hazard Analysis and Critical Control Points (HACCP). *EC Microbiology.* 16 (3): 1–7.

Janjic, M.J., Ciric, J., Boskovic, M. et al. (2017). Consumer responsibility for food safety. *Research in Agriculture.* 3 (1): 1.

Jevsnik, M., Helebec, V., and Raspor, P. (2008). Consumer interpretation of the term food safety. *Acta Alimentaria.* 37 (4): 437–448.

Jingar, S.C., Mehla, R.K., and Singh, M. (2014). Climatic effects on occurrence of clinical mastitisin different breeds of cows and buffaloes. *Archivos de Zootecnia.* 63: 473–482.

Johnson, R. (2014). Food fraud and economically motivated adulteration of food and food ingredients. *Congressional Research Service.* 1–40.

Kadjo, D., Ricker-Gilbert, J., Shively, G., and Abdoulaye, T. (2020). Food safety and adverse selection in rural maize markets. *Journal of Agricultural Economics. 71* (2): 412–438.

Kamboj, S., Gupta, N., Bandral, J.D. et al. (2020). Food safety and hygiene: a review. *International Journal of Chemical Studies.* 8 (2): 358–368.

Kamruzzaman, M. (2016). Food adulteration and authenticity. In: *Food Safety: Basic Concepts, Recent Issues and Future Challenges* (ed. J. Selamat and S.Z. Iqbal), 127–148. Switzerland: Springer.

Karanth, S., Benefo, E.O., Patra, D., and Pradhan, A.K. (2023). Importance of artificial intelligence in evaluating climate change and food safety risk. *Journal of Agriculture and Food Research.* 11: 100485.

Keener, K. (2009). *SSOP and GMP Practices and Programs.* Purdue Extension.

Kovats, R.S., Edwards, S.J., Hajat, S. et al. (2004). The effect of temperature on food poisoning: a time-series analysis of salmonellosis in ten European countries. *Epidemiology and Infection.* 132 (3): 443–453.

Krishna, A.K. and Govil, P.K. (2007). Soil contamination due to heavy metals from an industrial area of Surat, Gujarat, Western India. *Environmental Monitoring and Assessment.* 124: 263–275.

Kulinkina, A.V., Shinee, E., Herrador, B.R.G. et al. (2016). *The Situation of Water-related Infectious Diseases in the pan-European Region.* Geneva, Switzerland: World Health Organization.

Kunyanga, C., Imungi, J.K., and Okoth, M.W. (2011). Diversity and characteristics of supplementary foods sold and consumed by vulnerable groups in Kenya. *Journal of Applied Biosciences.* 45: 3019–3031.

Langhar, A.J., Ayers, T., Grass, J. et al. (2012). Nonpasteurized dairy products, disease outbreaks, and state laws- United States, 1993-2006. *Emerging Infectious Diseases.* 18 (3): 385–391.

Langiano, E., Ferrara, M., Lanni, L. et al. (2012). Food safety at home: knowledge and practices of consumer. *Journal of Public Health.* 20: 47–57.

Legesse, W. and Ambelu, A. (2004). *Lecture Notes: Personal Hygiene for Health Extension Workers.* Ethiopia Public Health Training Initiatives. USAID.

Li, C., Zhu, H., Li, C. et al. (2021). The present situation of pesticide residues in China and their removal and transformation during food processing. *Food Chemistry. 354*: 129552.

Linderhof, V., Thijs, L., and Stijn, R. (2021). The dilemmas of water quality and food security interactions in low- and middle-income countries. *Frontiers in Water.* 3: 736760.

Malakar, A., Snow, D.D., and Ray, C. (2019). Irrigation water quality—a contemporary perspective. *Water.* 11: 1482.

Manabe, S. (2019). Role of greenhouse gas in climate change**. *Tellus A: Dynamic Meteorology and Oceanography.* 71 (1): 1620078.

Marino, D.D. (2007). Water and food safety in the developing world: global implications for health and nutrition of infants and young children. *Journal of the Academy of Nutrition and Dietetics.* 107 (11): 1930–1934.

McCabe-Sellers, B.J. and Beattie, S.E. (2004). Food safety: emerging trends in foodborne illness surveillance and prevention. *Journal of American Dietetic Association.* 104 (11): 1708–1717.

Mead, P.S., Slutsker, L., Dietz, V. et al. (1999). Food-related illness and death in the United States. *Emerging Infectious Diseases. Centers for Disease Control and Prevention.* 5 (5): 607–625.

Melosi, M.V. (2004). *Garbage in the Cities: Refuse Reform and the Environment*. Pittsburgh: University of Pittsburgh Press.

Mihalache, O.A., Møretrø, T., Borda, D. et al. (2022). Kitchen layouts and consumers' food hygiene practices: ergonomics versus safety. *Food Control.* 131: 108433.

Miller, J.D., Frongillo, E.A., Weke, E. et al. (2021). Household water and food insecurity are positively associated with poor mental and physical health among adults living with HIV in western Kenya. *Journal of Nutrition* 151: 1656–1664.

Muntean, N.M., Jermini, I., Small, D. et al. (2003). Assessment of dietary exposure to some persistent organic pollutants in the Republic of Karakalpakstan of Uzbekistan. *Environmental Health Perspectives.* 111 (10): 1306–1311.

National Geographic. (2023). Food safety and security. https://education.nationalgeographic.org/resource/resource-library-food-safety-and-security/. (accessed 15 May 2023).

National Restaurant Association (NRA). (2023). Restaurant industry facts at a glance. https://restaurant.org/research-and-media/research/industry-statistics/national-statistics/ (accessed 5 November 2023).

Nerin, C., Contin, E., and Asensio, E. (2007). Kinetic migration studies using porapak as solid-food simulant to assess the safety of paper and board as food-packaging materials. *Analytical and Bioanalytical Chemistry.* 387 (6): 2283–2288.

Nerin, C., Aznar, M., and Carrizo, D. (2016). Food contamination during food process. *Trends in Food Science & Technology.* 48: 63–68.

Nji, Q.N., Babalola, O.O., and Mwanza, M. (2022). Aflatoxins in maize: can their occurrence be effectively managed in Africa in the face of climate change and food insecurity? *Toxins.* 14: 574.

Nounkeu, C.D. and Dharod, J.M. (2020). A qualitative examination of water access and related coping behaviors to understand its link to food insecurity among rural households in the West Region in Cameroon. *International Journal of Environmental Research and Public Health* 17: 4848.

Oliver, S.P., Boor, K.J., Murphy, S.C., and Murinda, S.E. (2009). Food safety hazards associated with consumption of raw milk. *Foodborne Pathogens and Disease.* 6 (7): 793–806.

Rahman, A., Tosepu, R., Karimuna, S.R. et al. (2018). Personal hygiene, sanitation and food safety knowledge of food workers at the university canteen in Indonesia. *Public Health of Indonesia.* 4 (4): 154–161.

Raspor, P. and Jevsik, M. (2008). Good nutritional practice from producer to consumer. *Critical Reviews in Food Science and Nutrition.* 48 (3): 276–292.

Raspor, P., Jevsnik, M., and Ambrozic, M. (2016). Food safety systems. In: *Food safety: Basic concepts, recent issues and future challenges* (ed. J. Selamat and S.Z. Iqbal), 3–31. Switzerland: Springer.

Rather, I.A., Koh, W.Y., Paek, W.K., and Lim, J. (2017). The sources of chemical contaminants in food and their health implications. *Frontiers in Pharmacology.* 8: 830.

Rezaei, A. (2018). Food safety: the farmer first health paradigm. *One Health.* 5: 69–73.

RSPH/DWI/Water (2017). Principles of Water Supply Hygiene. https://www.water.org.uk/principles-water-supply-hygiene (accessed 12 December 2023).

Saleem, M. and Umar, Z.N. (2013). Survey on the use of synthetic food colors in food samples procured from different educational institutes of Karachi City. *Journal of Tropical Life Science.* 3 (1): 1–7.

Scallan, E., Griffin, P.M., Angulo, F.J. et al. (2011). Foodborne illness acquired in the United States-unspecified agents. *Emerging Infectious Diseases.* 17 (1): 16–22.

Scheffler, R. (2009). Maximizing sanitation efforts in food processing: the importance of conveyor hygiene. *Trends in Food Science & Technology.* 20: 40–43.

Singh, P.K. (2014). *The Health and Economic Cost of Poor Sanitation.* WHO. https://www.who.int/southeastasia/news/speeches/detail/the-health-and-economic-cost-of-poor-sanitation (accessed 6 November 2023).

Sobsey, M.D. (2006). Drinking water and health research: a look to the future in the United States and globally. *Journal of Water and Health.* 4 (S1): 17–21.

Song, Q., Zheng, Y.-J., Xue, Y. et al. (2017). An evolutionary deep neural network for predicting morbidity of gastrointestinal infections by food contamination. *Neurocomputing.* 226: 16–22.

Spink, J., Moyer, D.C., and Whelan, P. (2016). The role of the public private partnership in food fraud prevention–includes implementing the strategy. *Current Opinions in Food Science.* 10: 68–75.

Spink, J., Bedard, B., Keogh, J. et al. (2019). International survey of food fraud and related terminology: preliminary results and discussion. *Journal of Food Science.* 84 (10): 2705–2718.

Talaviya, T., Shah, D., Patel, N. et al. (2020). Implementation of artificial intelligence in agriculture for optimisation of irrigation and application of pesticides and herbicides. *Artificial Intelligence in Agriculture.* 4: 58–73.

Tirado, M.C., Clarke, R., Jaykus, L.A. et al. (2010). Climate change and food safety: a review. *Food Research International.* 43: 1745–1765.

Tripathi, M., Khanna, S.K., and Das, M. (2007). Surveillance on use of synthetic colours in eatable vis a vis Prevention of Food Adulteration Act of India. *Food Control.* 18 (3): 211–219.

Tucker, M., Whaley, S.R., and Sharp, J. (2006). Consumer perceptions of food-related risks. *International Journal of Food Science & Technology.* 41 (2): 135–146.

Ucar, A., Yilmaz, M.V., and Cakiroglu, F.P. (2016). *Food safety-Problems and Solutions. Significance, Prevention and Control of Food Related Diseases.* Intechopen.

UN. (2019a). *World Population Prospects 2019: Highlights (ST/ESA/SER.A/423).* New York: Department of Economic and Social Affairs, Population Division, United Nations.

UN. (2019b). Goal 2: zero hunger . https://data.unicef.org/sdgs/goal-2-zero-hunger/#:~:text=Goal%202%20aims%20to%20end,of%20children%20and%20young%20people. (accessed 6 November 2023).

UN. (2020). Goal 6: ensure access to water and sanitation. https://www.un.org/sustainabledevelopment/water-and-sanitation/#:~:text=Worldwide%2C%20one%20in%20three%20people,people%20still%20practice%20open%20defecation (accessed 9 December 2020).

UN. (1948). Universal declaration of human rights. https://www.un.org/en/universal-declaration-human-rights/. (accessed 6 November, 2023).

Unneverhr, L.J. (2022). Addressing food safety challenges in rapidly developing food systems. *Agricultural Economics.* 53: 529–539.

Valoppi, F., Agustin, M., Abik, F. et al. (2021). Insight on current advances in food science and technology for feeding the world population. *Frontiers in Sustainable Food Systems.* 5: 626227.

Victoria, B.C. (2015). *Food Safety, Sanitation, and Personal Hygiene.* BC Cook Articulation Committee.

Vitali, A., Felici, A., Lees, A.M. et al. (2020). Heat load increases the risk of clinical mastitis in dairy cattle. *Journal of Dairy Science.* 103: 8378–8387.

Wang, W., Yu, Q., Meng, H. et al. (2018). Catalytic liquefaction of municipal sewage sludge over transition metal catalyst in ethanol-water co-solvent. *Bioresource Technology.* 249: 361–367.

WASH Advocates. (2013). Global Food Security: The Critical Role of Water, Sanitation and Hygiene (WASH). WASH Advocates, Washington D.C., USA.USA

Waturangi, D.E., Wennars, M., Suhartono, M.X., and Wijaya, Y.F. (2013). Edible ice in Jakarta, Indonesia, is contaminated with multidrug-resistant vibrio cholerae with virulence potential. *Journal of Medical Microbiology* 62: 352–359.

WFP. (2019). Covid-19 will double number people facing food crises unless swift action taken. https://www.wfp.org/news/covid-19-will-double-number-people-facing-food-crises-unless-swift-action-taken (accessed 6 November 2023)

WHO. (2003). *Several Foodborne Diseases are Increasing in Europe*. Geneva, Switzerland: WHO.

WHO. (2005). *Sanitation and Hygiene Promotion: Programming Guidance*. Geneva, Switzerland: WHO.

WHO. (2006). *Five Keys to Safer Food Manual*. Geneva, Switzerland: WHO.

WHO. (2015). *Progress on Sanitation and Drinking Water: 2015 Update and MDG Assessment*. Geneva, Switzerland: WHO.

WHO. (2016).Technical consultation on the implementation of the regional framework for sound management of public health pesticides, 2016–2020. Document WHO EM/MAL/382/E.

WHO. (2018). Arsenic [Fact sheet].https://www.who.int/news-room/fact-sheets/detail/arsenic.(accessed 6 November 2023)

WHO. (2019). *Water, Sanitation, Hygiene and Health: A Primer for Health Professionals*. Geneva, Switzerland: WHO.

Woldt, M. and Moy, G.G. (2015). *Literature Review on Effective Food Hygiene Interventions for Households in Developing Countries*. USDA.

Xi, X., Wei, S., Lin, K.L. et al. (2021). Digital technology, knowledge level, and food safety governance: implications for national healthcare system. *Front Public Health*. 9: 753950.

Zhou, X., Li, Z., Zheng, T. et al. (2018). Review of global sanitation development. *Environment International*. 120: 246–261.

Index

Please note: entries in *italics* refer to separate figures, whereas **bold** entries refer to tables.

Food Security in the Developing World, First Edition. Edited by Muhammad Khalid Bashir, Steven G.M. Schilizzi, and Ghaffar Ali.

c

d